TRANSLATION

TRADUZIONE
AUTOMATICA
TRADUCCIÓN
AUTOMÁTICA
KONEKÄÄNTÄMINEN
机器翻译
МАШИННЫЙ
ПЕРЕВОД
기계 번역
TRADUCTION
AUTOMATIQUE

The MIT Press Essential Knowledge Series

MACHINE TRANSLATION

THIERRY POIBEAU

The MIT Press | Cambridge, Massachusetts | London, England

This book was set in Chaparral Pro by Toppan Best-set Premedia Limited. Printed and bound in the United States of America.

Library of Congress Cataloging-in-Publication Data is available.

ISBN: 978-0-262-53421-5

10   9   8   7   6   5   4   3   2   1

CONTENTS

SERIES FOREWORD

The MIT Press Essential Knowledge series offers accessible, concise, beautifully produced pocket-size books on topics of current interest. Written by leading thinkers, the books in this series deliver expert overviews of subjects that range from the cultural and the historical to the scientific and the technical.

In today's era of instant information gratification, we have ready access to opinions, rationalizations, and superficial descriptions. Much harder to come by is the foundational knowledge that informs a principled understanding of the world. Essential Knowledge books fill that need. Synthesizing specialized subject matter for nonspecialists and engaging critical topics through fundamentals, each of these compact volumes offers readers a point of access to complex ideas.

Bruce Tidor
Professor of Biological Engineering and Computer Science
Massachusetts Institute of Technology

ACKNOWLEDGMENTS

This book would not have been possible without the support of colleagues and friends. I want to thank Michelle Bruni, Elizabeth Rowley-Jolivet, Pablo Ruiz Fabo, and Bernard Victorri for their help during the preparation of this book. My gratitude also goes to the editorial and production staff at MIT Press, particularly Marie Lufkin Lee and Katherine A. Almeida. Finally, I want to thank the anonymous reviewers for their careful reading and their many insightful comments and suggestions.

Thierry Poibeau is a member of LATTICE, a research laboratory supported by CNRS, Ecole normale supérieure (ENS), PSL Research University, Université Sorbonne nouvelle, and USPC.

INTRODUCTION

In Douglas Adams' humorous saga *The Hitchhiker's Guide to the Galaxy*,[1] all one needs to do to understand any language is to introduce a small fish (the Babel fish) into one's ear. This improbable invention is of course related to the idea of a universal translation device,[2] and more generally to the key problem of language diversity and comprehension. The name of the fish is a transparent allusion to the Biblical episode of Babel, when God scrambled language so that humans could no longer understand one another.

A significant number of thinkers, philosophers, and linguists—and, more recently, computer scientists, mathematicians, and engineers—have tackled the question of language diversity. Moreover, they have imagined theories and devices intended to solve the problems caused by this diversity. Since the advent of computers (after the Second World War), this research program has materialized through the design of *machine translation* tools—in other words, computer programs capable of automatically producing in a target language the translation of a text in a source language.

This research program is very ambitious: it is even one of the most fundamental in the field of artificial intelligence. The analysis of languages cannot be separated from the analysis of knowledge and reasoning, which explains the interest in this field shown by philosophers and specialists of artificial intelligence as well as the cognitive sciences. This brings to mind the test proposed by Turing[3] in 1950: the test is successfully completed if a person dialoguing (through a screen) with a computer is unable to say whether her discussion partner is a computer or a human being. This test is foundational, because developing an operational conversational agent presupposes not only understanding what the discussion partner says (at least to some extent), but also inferring from what has been said a relevant utterance that helps the whole conversation move forward. For Turing, if the test is successful, it means that the machine has a certain degree of intelligence. This question has fueled considerable debate, but we can at least agree on the fact that a robust conversational system would involve formalizing some mechanisms of understanding and reasoning.

Machine translation involves different processes that make it at least as challenging as developing an automatic

The analysis of languages cannot be separated from the analysis of knowledge and reasoning, which explains the interest shown by philosophers and specialists of artificial intelligence as well as cognitive sciences in [machine translation].

dialoguing system. The degree of "understanding" shown by the machine can be very partial: for example, the Eliza system developed by Weizenbaum in 1966 was able to simulate a dialogue between a psychotherapist and his patient. The system in fact just derived questions from the patient's utterances (for example, the system was able to produce the question "why are you afraid of X?" from the sentence "I am afraid of X"). The system also included a series of ready-made sentences that were used when no predefined patterns seemed to be applicable (for example "could you specify what you have in mind?" or "really?"). Despite its simplicity, Eliza had great success, and some patients really thought they were conversing with a real doctor through a computer.

The situation is completely different when considering machine translation. Translation requires in-depth understanding of the text to be translated. Moreover, transposition into another language is a delicate and difficult process, even with news or technical texts. The aim of machine translation is not, of course, to address literature or poetry; rather, the idea is to give the most accurate translation of everyday texts. Even so, the task is immensely difficult, and current systems are still far from satisfactory.

However, and despite its limitations, from a more theoretical point of view, machine translation also makes us take a fresh look at old and widely investigated questions:

What does it mean to translate? What kind of knowledge is involved in the translation process? How can we transpose a text from one language to another? These are some of the questions that are addressed in this book.

This short book aims at providing an overview of the progress in machine translation since the Second World War. Some pioneers will be mentioned, but it is mainly the research implemented with computers that will be addressed. The content of the book is thus partly historical, since the main approaches to the problem will be presented in an intuitive manner: the idea is to make sure that the reader can understand the main principles without having to know all the technical details. Specifically, recent approaches based on the statistical analysis of very large corpora of texts will be presented, but these approaches are highly technical and we will skip the mathematical details that are not necessary to grasp the overall idea. More technical books exist for those who are interested in the full details of the different approaches.

The book begins with a presentation of the main problems one has to solve when developing a machine translation system (chapter 2). The journey continues with a quick overview of the evolution of machine translation (chapter 3), followed by a more detailed presentation of the history of the field, from its beginnings before the advent of computers (chapter 4) to the most recent advances based on deep learning (chapter 12). Along the way, we will encounter all the main approaches developed since the field's beginning: rule-based approaches (chapter 5) up to the ALPAC report and its consequences (chapter 6); and the advent of parallel corpora (chapter 7), which fueled research in the field after the 1980s, first through the example-based paradigm (chapter 8), then through the most popular statistical paradigm (chapter 9) along with its more recent developments—the segment-based approach (chapter 10) and the introduction of more linguistic knowledge to the systems (chapter 11). This book is not limited to a presentation of the main approaches to the problem: we will also address evaluation issues (chapter 13), which can be either manual or automatic, and the closing chapter will give some details about the commercial situation of the field as well as its main actors worldwide (chapter 14). Although the domain is evolving quickly, including from a commercial point of view, we think it is important to address industrial issues since machine translation is now a key technology for several prominent domains, from defense to media and telecommunications. Lastly, we conclude with some observations on the current state of the field (chapter 15) and provide some references for further reading.

THE TROUBLE WITH TRANSLATION

Before addressing machine translation, it is important to investigate the notion of translation in itself. How do we proceed when we translate? What makes a translation a good translation? In the course of this chapter, we will see that these questions are hard to answer and have already given rise to an abundant literature. In the second part of this chapter, we will investigate why understanding a sentence—something that is easy and natural for humans—is one of the most difficult things to do with computers, despite their incredible calculation power.

What Does It Mean to Translate?

The answer to this question may seem obvious: to translate is to transpose a source-language text into a target-language text. However, one can easily see that this deceptively simple answer refers in fact to a dramatically complex problem. What does it mean to "transpose a text"? How do we go from a source language to a target language? How does one find equivalent expressions between two languages? Should the translation be based on words, chunks of words, or even sentences? And, more fundamentally, how can one determine what the meaning of a text or an expression is? Does everybody have the same understanding of a text? If not, how can this issue be handled in the translation process?

As should be clear from the previous paragraph, translation is connected with a large number of questions dealing with linguistics, but also psychology or even philosophy when the nature of meaning is at stake. Instead of addressing these highly complex questions (with no clear answer!), it is probably more useful to take a step to the side and try to determine what the characteristics of a "good" translation are.

What Is a Good Translation?

A first crucial issue when addressing translation is that no one knows how to formally define what constitutes a "good" translation. We should thus not expect to make

A crucial issue when addressing translation is that no one knows how to formally define what constitutes a "good" translation.

much headway from this perspective, but at least some criteria can be found in the literature.

Criteria for a "Good" Translation

The translation of a text should be faithful to the original text: it should respect the main characteristics of the original text, the tone and style, the details of the ideas as well as its overall structure. The result should be easy to read in the target language, and it should also be linguistically correct, which means that a subtle process of reformulation must be used. Ideally, the reader should not realize he is reading a translation if he does not know the origin of the text, which implies that all formulaic and idiomatic expressions should be rendered appropriately.

As a result, the translator must perfectly understand the text he has to translate, but he must also have an even better knowledge of the target language. This is the reason why professional translators usually only translate into their mother language so that they have a perfect understanding and knowledge of the expressions to be used to render the source text accurately.

The inherent subjectivity of these criteria is undeniable, however. What is considered as a "good" translation by some readers may be a bad one according to another person. This situation frequently crops up when professional translators work with authors they are not familiar with or when the translator does not know in what context his translation will be used.

What is expected of a translation can vary radically depending on the clients, the era, the nature of the text, its usage, or even context. Technical texts are not translated in the same way as literary texts. A specific adaptation of the original text is necessary when the text concerns a world that is remote from the world of the reader in the target language (for example, if a Japanese text from the twelfth century is translated into modern English). The translator has to choose between staying close to the original text or making use of paraphrasing to ensure comprehension (especially with historical contexts, unfamiliar events, etc.). The tone and the style of a text are also highly subjective notions that are largely related to the language under consideration.

As one can easily see from this quick overview, all these subjective features make the evaluation of the task a difficult problem.

Some pitfalls are, however, well known and frequently addressed in the literature on the topic. Word-for-word translation is not a good practice, since the result is often hard to understand and not idiomatic in the target language. Deceptive cognates and syntactic duplicates should, of course, be banished since they lead to nonsense (the French word "*achèvement*" should be translated as "completion" in English, and not as "achievement," for example). It is also well known that a translator should first read the whole text, or at least a large part of the text, to be translated so as to avoid local mistranslations. A good knowledge of the clients, context, and future use of the translated text can also help to adjust the translation to the target.

Consequences for Machine Translation

From what we have seen so far, it is clear that translation is a complex process involving high-level cognitive and linguistic capabilities. A translator must be at ease with the two languages involved, and he must have special skills to reformulate a source language in a target language that does not have the same wording or the same structure.

These kinds of skills are not directly available to machines. Artificial systems are still in their infancy from this point of view and are very far from the capacities of a human being when it comes to reasoning, inferring, and reformulating. To be able to reformulate a sentence, one must of course have a good command of the language itself, but one must also master the search for analogy between concepts, which is much more complicated that just equivalencies between words and expressions.

Developers of artificial systems are aware of these limitations. Very few researchers have tried to develop machine translation systems for literary texts: nearly everybody agrees that machine translation is a difficult task that is far from being resolved, and that only mundane texts (e.g., news, technical texts) should be addressed. The idea is not to replace human translators who are the only ones able to translate novels or poetry. Even technical texts pose specific difficulties since they employ a very technical vocabulary that has first to be introduced into the system in order to obtain relevant translations. The goal of machine translation is now considered mainly to be that of providing the user with some help and, in some professional contexts, enabling him to decide whether a human translator needs to be called on or not.

The overall quality achievable by machine translation has also been a matter of much debate. The ultimate goal is to obtain a quality of translation equivalent to that of a human being. People agree that this is highly challenging and also hard to formalize, since the quality of a translation is related to the nature and complexity of the text to be translated.

For a long time, machine translation used local techniques that could be compared, to a certain extent, to a word-for-word translation process, even if most systems now also take more complex expressions into consideration. Information at the text level is rarely taken into account, even though it is well known that the text can provide important information for the translation process. The tonality or the style of a text, for example, is always ignored: this kind of information is in fact too hard to formalize for automatic systems.

In a way, even the sentence level is too complex for most current systems. It is generally assumed that these systems perform a sentence-by-sentence translation, which is true to a certain extent, but the translation process generally involves, in fact, fragments of sentences.[1] The translation of a full sentence then consists in assembling the translations of these local fragments. It is therefore not surprising that machine translation sometimes provides strange results and quite often utter nonsense. Morphology (the analysis of the structure of words) and syntax (the analysis of the structure of sentences) are rarely taken into account, and this has particularly dramatic consequences for some languages. For example, some are said to be highly inflectional, which means that word forms can change depending on the grammatical function of the word in the sentence (subject, complement, etc.). In this context, it is clear that an automatic process will not be able to provide the right word form in the target language without a proper syntactic analysis (i.e., an analysis of the relative grammatical function of the different words in the sentence).

Last but not least, one should understand why processing languages with computers is difficult, even when dealing with easier tasks than machine translation. A language has thousands of words, with different surface forms ("*to dance*," "*danced*," "*dancing*"), different meanings, and different structures. Compounds (e.g., "*round table*," which generally designates an event and not an object), light verbs (e.g., "*to take a shower*," where "*take*" has little semantic content), and idioms or frozen expressions (e.g., "*kick the bucket*," the meaning of which has nothing to do with "kick" or "bucket") make the task even more complex, since it is then necessary to spot complex expressions and not only isolated words. The following section aims at showing some of the issues at stake.

Why Is It Difficult to Analyze Natural Language with Computers?

Apart from the lack of information on the client, the context, or the style of the text under consideration for translation, the main issue is related to the task itself. Processing natural languages (as opposed to processing formal languages, such as the programming languages used by computers) is difficult in itself, mainly because at the heart of natural language lie vagueness and ambiguity.

Natural Languages and Ambiguity

Linguists as well as computer scientists have been interested ever since the creation of computers in natural language processing, a field also called computational linguistics. Natural language processing is difficult because, by default, computers do not have any knowledge of what a language is. It is thus necessary to specify the definition of a word, a phrase, and a sentence. So far, things may not seem too difficult (however, think about expressions like: *"isn't it," "won't," "U.S.," "$80"*: it is not always clear what is a word and how many words are involved in such expressions) and not so different from formal languages, which are also made of words. The main difference lies in the fact that every word and every expression of a given natural language can be ambiguous.

Let's take some famous examples such as *"the chicken is ready to eat"* or *"there was not a single man at the party."* These are textbook examples and may seem a bit far-fetched. However, they illustrate some well-known problems in language processing: in the first example, should one give the chicken something to eat, or is it the chicken that is ready to be eaten? In the second example, does the speaker mean that there were no men at the party, or does he mean that all the men there were married? These sophisticated examples should not mask the fact that ambiguity is in fact pervasive and is also part of the most mundane words and expressions. Just in these two examples, we can remark that *"chicken"* can refer to an animal or a kind of meat, but also a coward. A party can designate (according to Wordnet[2]) *"an organization to gain political power," "a group of people gathered together for pleasure," "a band of people associated temporarily in some activity,*" "*an occasion on which people can assemble for social interaction and entertainment,*" or even "*a person involved in legal proceedings*"). "*Party*" can also be a verb for "*have or participate in a party,*" etc.

One answer to this problem is just to record all these different meanings in a dictionary, and this in a way already exists since we mentioned, for example, Wordnet, a lexical database that can be used by humans as well as by computers. However, one quickly realizes that this is not a working solution, since once all these meanings have been stored in the dictionary, the problem is then to find a way to choose the right meaning of each occurrence (that is to say, for each word used in context).

A normal dictionary usually contains around 50,000 to 100,000 entries (i.e., different words) that can in turn generate more surface forms, or words as they are found in texts. For example, "*texts*" is not a dictionary entry, since it is just a surface form of the word "*text*" ("*texts*" is the plural of "*text*," and this is supposed to be known by the end user). This point of departure is assumed by nearly all dictionaries made for normal human perusal. In a dictionary, only the singular is stored for nouns and adjectives, and the infinitive for verbs; the dictionary form of a word is usually called a lemma. In English, the number of surface forms is limited, but the problem is worse for a language like French. For other languages like Finnish, the theoretical number of surface forms is huge and could even be considered infinite, since the language has at least 12 cases and lots of suffixes and particles that can be combined in various ways. Trying to store all these forms in a dictionary is probably not a good idea!

What makes things even more difficult is that to decide on the meaning of a word or an expression (does "*party*" here mean "*an organization to gain political power*" or "*a group of people gathered together for pleasure*"?), one has to take context into account. But the context itself is generally ambiguous, leading to a potentially unresolvable problem. Moreover, it has been demonstrated that word senses are not mutually exclusive and that "word usages often fall between dictionary definitions" (Kilgarriff, 2006). This is one of the main consequences of the pervasive vagueness of languages.

What may seem paradoxical is that humans, who cannot process numbers as fast or as accurately as computers, are in fact very good at handling these kinds of problems. Most of us do not see any ambiguity in most sentences, even when there are thousands of meanings that could possibly be considered. This aspect of language complexity was simply not grasped by most of the early researchers in the domain or, to be more exact, this complexity was largely underestimated.

The way language is processed (and more specifically the way an utterance is understood) remains largely obscure, even nowadays in the era of neuroimaging. Understanding seems to be natural, direct, and largely unconscious. It is highly doubtful that all possibilities are considered in order to obtain a semantic representation of a sentence. Thanks to the communication context, the brain probably directly activates the "right" meaning, without even considering alternate solutions. A parallel has sometimes been proposed with the Necker cube, the representation of a cube seen in perspective with no depth cue (figure 1).

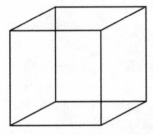

Figure 1   The Necker cube, the famous optical illusion published by Louis Albert Necker in 1832. (Image licensed under CC BY-SA 3.0 via Wikimedia Commons. From https://commons.wikimedia.org/wiki/File:Necker _cube.svg.)

The drawing is "ambiguous" in that no cue makes it possible to determine which side of the cube is in front and which side is at the back. However, it was noticed by Necker (and others before him) that humans naturally select one of the representations so that it makes sense and is coherent with the image of a cube in nature. Both interpretations, that is to say two different cubes, can be seen alternately, but they cannot both be considered simultaneously, since this would violate predefined conceptions embedded in the brain. One can also think about Escher's asymmetrical drawings that take advantage of quirks of perception and perspective: these images are largely based on representations that violate our pre-conceived representations of space.

These examples should remind us that the brain is able to interpret (and sometimes correct) perceptions in accordance with predefined schemas. Without going into too many details, this theory is also in line with the notion of *Gestalt*, which refers to the idea that the brain interprets a whole from its parts and a part from the whole. Applied to language, this means that the meaning of a word is largely determined by the larger context, which itself depends on the meaning of the words it is composed of. There is a dynamic co-construction of interpretation in the brain that is absolutely natural and unconscious.[3]

Consequences for Machine Translation

We have seen in this chapter that the main issue for natural language processing is ambiguity: it is simply difficult to determine the meaning of a word. Meaning depends on context, but the notion of context is itself also vague and ambiguous.

We should add that determining the number of meanings per word (what is generally called word sense) is also an open issue, since from one dictionary to another, the number of word senses differs: some dictionaries are more precise and contain a more fine-grained description of meaning, while others prefer to limit the number of senses per word, depending on their conceptual choices and on their intended readership.

Despite these issues, it has been generally assumed that in order to produce high-quality translations, one must first provide a precise and accurate description of the meaning of sentences. Advances in machine translation were then linked to the progress made in text understanding, which largely drove the field for several years, as we will see in the following chapters. However, these hypotheses are called into question: statistical approaches can make use of large quantities of text available on the web and calculate possible equivalencies between language without using any predefined dictionaries or high-level formalisms. In subsequent chapters (see, especially, chapters 9 to 12), we will discuss how accurate these models are and to what extent they avoid (or integrate) semantic information.

Artificial and Natural Systems

A much-debated question in the field of machine translation is the extent to which artificial systems should reproduce the strategies used by humans for translating. In other words, can we learn something from observing the working practices of professional translators?

This is another hard question, and the first thing to stress is that we do not know much about the cognitive processes involved in the translation task. Moreover, translation strategies probably vary largely from one translator to another. It is clear that translating requires going beyond the simple word-for-word approach, as we have already seen, but it is debatable whether professional translators systematically do a deep syntactic and semantic analysis of the sentence to be translated. It is, for example, clear that professional interpreters (doing on-the-fly speech translation) often translate semiautonomous groups of words without having heard the full sentence, especially in the case of a long sentence.

This approach can be compared to that of statistical systems that do not perform a deep analysis of the sentence to be translated, but identify groups of words that work together. The parallel is not completely exact, since interpreters always select groups of words that are relatively autonomous in the sentence (generally full phrases), whereas a statistical analysis will extract any regular group of words without relying too much on syntactic constraints. However, as already said, statistical systems are very good at recognizing multiword expressions (compounds, idioms, etc.) that are perceived as single units even by humans, as the psychoanalysis of language shows.

Recent experiments have reinforced this point of view, since even high-level syntactic structures can correspond to regular patterns. These structures are sometimes called "constructions" (specific syntactico-semantic structures registered as such in our brain) or "prefabs" (like a home made of prefabricated elements that can be quickly assembled to obtain a modular construction). In this framework, syntax is not as prominent as in traditional approaches: the sentence is seen as an assemblage of "prefab units," or, put differently, an assemblage of complex sequences stored as such in the brain. The analysis is thus simpler, since, if this hypothesis is correct, the brain does not really have to take into account each individual word but has direct access to higher-level units, reducing both the overall ambiguity and the complexity of the sentence-understanding process.

Hence, it is not certain that interlingual systems, based on a complete understanding and abstract representation of sentences, are the most realistic ones from a cognitive point of view, contrary to what has long been believed. We will come back to these issues once we have described in detail the different approaches considered in the domain.

A QUICK OVERVIEW OF THE EVOLUTION OF MACHINE TRANSLATION

In this chapter we examine the different possible approaches and the main tendencies observed in the domain of machine translation since its beginnings. It is important to have an idea of the main challenges and the main evolutions of the domain before diving into more detail. Each of these approaches will then be detailed in the following chapters.

Rule-Based Systems: From Direct to Interlingual Approaches

Different approaches and different techniques have been used for machine translation. For example, translation can be direct, from one language to the other (i.e., with no intermediate representation), or indirect, when a system

Translation can be direct, from one language to the other (i.e., with no intermediate representation), or indirect, when a system first tries to determine a more abstract representation of the content to be translated.

first tries to determine a more abstract representation of the content to be translated. This intermediate representation can also be language independent so as to make it possible to directly translate one source text into different target languages.

Each system is unique and implements a more or less original approach to the problem. However, for the sake of clarity and simplicity, the different approaches can be grouped into three different categories, as most textbooks on the topic do.

1. A *direct translation system* is a system that tries to produce a translation directly from a source language to a target language with no intermediate representation. These systems are generally dictionary-based: a dictionary provides a word-for-word translation, and then more or less sophisticated rules try to re-order the target words so as to get a word order as close as possible to what is required by the target language. There is no syntactic analysis in this kind of system, and reordering rules apply directly to surface forms.

2. *Transfer systems* are more complex than direct translation systems, since they integrate some kind of syntactic analysis. The translation process is then able to exploit the structure of the source sentence provided by the syntactic analysis component, avoiding the word-for-word

limitation of direct translation. The result is thus supposed to be more idiomatic than with direct translation, as long as the syntactic component provides accurate information on the source and on the target language.

3.   The most ambitious systems are based on an *interlingua*, which is a more or less formal representation of the content to be translated. Extensive research has been done on the notion of interlingua. Fundamental questions immediately arose such as: how deep and precise should an interlingua be so as to provide a sound representation of the sentence to be translated? Instead of developing a completely artificial language, which is known to be a very complex task, English is often used as an interlingua, but this is in fact quite misleading, since the representation is then neither formal nor language-independent. It is thus better to speak of a "pivot language," or simply a "pivot," when the interlingua is a specific natural language (English, in most cases, as we have just seen, but Esperanto and other languages have also been used in the past). In this context, when translating from language A to language B, the system first tries to transfer the content of A to the pivot language before translating from the pivot to the target language B.

These three kinds of approaches can be considered to form a continuum, going from a strategy that is very close to the surface of the text (a word-for-word translation) up to systems trying to develop a fully artificial and abstract representation independent of any language. These varying strategies have been summarized in a very striking figure called the "Vauquois triangle," from the name of a famous French researcher in machine translation in the 1960s (figure 2).

Direct transfer, represented at the bottom of the triangle, corresponds to word-for-word translation. In this

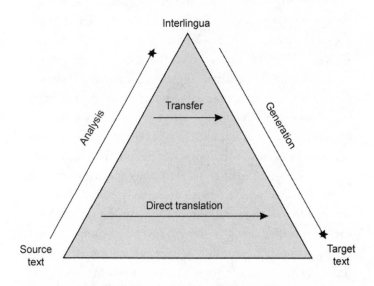

Figure 2 Vauquois' triangle (image licensed under CC BY-SA 3.0, via WikiMedia Commons). Source: https://en.wikipedia.org/wiki/File:Direct_translation_and_transfer_translation_pyramind.svg.

framework, there is no need to analyze the source text and, in the simplest case, a simple bilingual dictionary is enough. Of course, this strategy does not work very well, since every language has its own specificities and everybody knows that word-for-word translation is a bad strategy that should be avoided. It can nevertheless give some rough information on the content of a text and may seem acceptable when the two languages considered are very close (same language family, similar syntax, etc.).

Researchers have from the very beginning also tried to develop more sophisticated strategies to take into account the structure of the languages at stake. The notion of "transfer rules" appeared in the 1950s: to go from a source language to a target language, one needs to have information on how to translate groups of words that form a linguistic unit (an idiom or even a phrase). The structure of sentences is too variable to be taken into account directly as a whole, but sentences can be split into fragments (or chunks) that can be translated using specific rules. For example, adjectives in French are usually placed after the noun, whereas they are before the noun in English. This can be specified using transfer rules. More complex rules can apply to structures like *"je veux qu'il vienne"* ⇔ *"I want him to come,"* where there is no exact word-for-word correspondence between the two sentences (*"I want that he comes"* is not very good English, and *"je veux lui de venir"* is simply ungrammatical in French).

The notion of transfer can also be applied to the semantic level in order to choose the right meaning of a word depending on the context (for example, to know whether a given occurrence of "*bank*" refers to the bank of a river or to a money-lending institution). In practice, this is a hard problem if done manually, since it is impossible to predict all the contexts of use of a given word. For exactly the same reason, this quickly proved to be one of the most difficult problems to solve during the early days of machine translation. We will see in the following chapters that, more recently, statistical techniques have produced much more satisfactory results, since the problem can be accurately approached by the observation of very large quantities of data—the kind of thing computers are very good at, and humans less so (at least when they try to provide an explicit and formal model of the problem).

Last but not least, another family of systems is based on the notion of interlingua, as we have already seen in the previous section. Transfer rules, by definition, always concern two different languages (i.e., English to French in our examples so far) and thus need to be adapted for each new couple of languages considered. The notion of interlingua is supposed to solve this problem by providing a language-independent level of representation. Compared to transfer systems, the interlingual approach still needs one analysis component to go from the source text to the interlingual representation, but then this representation can give birth to translations into several languages directly. The production of a target text from the interlingual representation format, however, requires what is called a "generation module"—in other words, a module able to go from a more or less abstract representation in the interlingual format to linguistically valid sentences in the different target languages.

Interlingual systems are very ambitious, since they need both a complete understanding of the sentence to be translated and accurate generation components to produce linguistically valid sentences in the different target languages. Moreover, we saw in the previous chapter that understanding text is to a great extent an abstract notion: what does it mean to "understand"? What information is required so as to be able to translate? To what extent is it possible to formalize the comprehension process given the current state of the art in the domain? As a result, and despite several years of research by several very active groups, interlingual systems have never been deployed on a very large scale. The issues are too complex: understanding a text may potentially mean representing an infinity of expressed and inferred information, which is of course highly challenging and simply goes beyond the current state of the art.

The Revolution of Statistical Machine Translation Systems

The classification of machine translation systems provided in the previous section is challenged by new approaches that have appeared since the early 1990s. The availability of huge quantities of text, especially on the Internet, and the development of the capacity of computers have revolutionized the domain.

Most current industrial machine translation systems, and especially the most popular ones (Google translation, Bing translation) are based on a statistical approach that does not completely fit into the previous classification. These systems are not primarily based on large bilingual dictionaries and sets of hand-crafted rules. The first statistical systems implemented a kind of direct translation approach, since they tried to find word equivalences between two different languages by directly looking at very large amounts of bilingual data (initially coming from specialized international institutions and, more recently, for the most part, harvested on the web).

Statistical approaches are now considerably more precise. They no longer deal with isolated words but are now able to spot sequences of words (such as compounds, idioms, frozen expressions, or just regular sequences of several words) that need to be translated as a whole. The most recent approaches even try to tackle the problem directly at the sentence level. It should be noted that these systems have their own internal representations that are generally not directly readable by a human being. It is thus necessary to consider the nature of these representations: to what extent do they render semantic information? Can we draw any parallel between this approach and the way humans handle language?

The success of these systems lies in the fact that they are able to grasp regular equivalencies between languages, but also more or less frozen sequences in a text, based on a purely statistical analysis. Since, as we have seen, meaning is not something formally defined but corresponds to the way words are used, a purely statistical approach can be quite powerful in discovering regularities and specific contexts of use. However, for a long time (and in most available online systems still now) equivalencies were calculated at a local level and involved fragments of text that often overlap. The main difficulty was then to make sense of all these fragments and equivalencies: statistical systems then had to deal with a multiplicity of fragments that had to be assembled to give birth to a well-formed sentence. These fragments may provide contradictory information and are thus not fully compatible. The task could be compared to assembling a jigsaw puzzle using a set of pieces coming from 10 different puzzles. In chapter 12, we give a brief overview of the most recent approaches based on deep learning, a new and again completely different paradigm, which attempts to tackle the problem directly at the sentence level. Deep learning approaches thus have the potential to give considerably more precise results.

A Quick Historical Overview

The history of machine translation can be summarized as follows:

• Until the 1940s, some researchers thought about the problem of the automation of translation, but these pioneering studies were not taken up since it was not possible to develop and test them in practice.

• From the 1940s to the mid-1960s, with the advent of the first computers, several teams developed operational machine translation systems. Expectations concerning the development of the field were high. The approaches used were sometimes rather naïve, but some researchers already imagined more fundamental approaches, some of which were seminal and rediscovered some years later.

• In 1965–1966, the ALPAC report, commissioned by American funding agencies, had a dramatic impact on the field. The conclusions of the report were highly negative, describing the research done up to then as flawed and useless. Funding that was already dramatically shrinking gradually petered out. However, one should note that the report also stressed the need to develop more fundamental research combining computers and linguistics, so as to allow the field to make more progress in text understanding (and more fundamentally in parsing and in automatic semantic analysis).

• The following years, up until the end of the 1980s, were not very productive for machine translation, especially in the Anglo-American world. New groups nevertheless emerged in Europe and other countries. On the other hand, research in computational linguistics was blooming during the same period for speech as well as for written text: the 1960s and 1970s saw major developments in parsing (automatic syntactic analysis), semantics, and text understanding for example, as suggested in the 1966 ALPAC report (see chapter 5).

• The 1990s saw the advent of a new approach based on statistics and very large bilingual corpora. This trend clearly derived from a series of seminal papers written by a research group working at IBM in the late 1980s and early 1990s. These papers had a considerable impact, along with the development of statistical and empirical approaches in natural language processing. The most popular translation systems nowadays (Google and Bing translation) are all based on a variant of this approach.

•   Recently, the very high level of demand for automatic translation over the web has also had the effect of reinstating machine translation at the heart of the field of computational linguistics, after several decades in purgatory. A new approach based on deep learning is also completely revolutionizing the field since the mid-2010s.

We now need to examine these different stages in more detail, so as to better understand the approaches and their main challenges as well as the limitations of each system.

BEFORE THE ADVENT
OF COMPUTERS...

Machine translation is closely related to the advent of computers, allowing scientists to imagine a fully automatic translation process. At the same time, we should not forget that very early on, there were also philosophical, religious, and scholarly speculations about the possibility of automating translation that are important for the history of the field. Eventually, the first half of the twentieth century saw the design of prototypes that prefigured some of the systems developed since the 1950s.

The Question of Universal Languages

A long and relevant tradition from this perspective is the quest for a universal language. If such a language were to exist, it would by nature eliminate the need for translation, given its universal nature. More realistically, one could think of an artificial language that would facilitate translation between existing languages. Needless to say, this is a key point for machine translation. Different groups have explored this idea of a universal language; the systems developed on this basis are commonly called "interlingual," as seen in the previous chapter.

A Long Tradition

In the Western tradition, the Ancients often refer to an Adamic language, that is, a hypothetical and universal protolanguage spoken by humanity before the story of Babel. This aroused the interest of Leibniz, even if he was also among the first to partially abandon this tradition, since he believed it was impossible to rediscover this Adamic language from our modern languages. He nevertheless developed a project intended to eliminate ambiguity in languages, by defining a new artificial language that no longer referred to any supposed Adamic tradition, for the purpose of solving various problems, such as moral, legal, or philosophical dilemmas (Leibniz, 1951).

Descartes[1] is often cited along with Leibniz in this respect, since he was also interested in the idea of a universal language and its relationship with existing languages. Witness the following passage about a proposal for a universal language: "If [someone] put into [his] dictionary a single symbol corresponding to *aymer*, *amare*, *philein* and each of the synonyms, a book written in such symbols could be translated by all who possessed the dictionary" (Descartes, letter to Mersenne on November 20, 1629). This passage greatly inspired machine translation pioneers, since Descartes' proposal aimed to replace words with unambiguous codes ("symbols" corresponding to numerical codes that are independent from the languages considered; symbols replace words in Descartes' proposal).

In the wake of these proposals, several attempts to develop a "numerical dictionary" emerged at the end of the seventeenth century in Europe. A numerical dictionary is a dictionary in which a specific number (an identifier) is associated with each word or concept. Those who attempted the task include Cave Beck in 1657, Johann Joachim Becher in 1661, Athanasius Kircher in 1663, and John Wilkins in 1668. Hutchins[2] mentions that Becher's dictionary was republished in Germany in 1962 as "On mechanical translation: A coding attempt from 1661."[3] Also worth mentioning in France are Joseph de Maimieux (inventor of the term pasigraphy in 1797, which refers to a style of writing, or universal notation system) and Arman-Charles-Daniel de Firmas-Périés, who developed such a system in 1811. The main application was the encoding and decoding of messages, essentially for military needs.

However, it is important to refrain from seeing these initiatives as direct precursors of machine translation. Leibniz's and Descartes' essentially aimed to solve philosophical, logical, and moral problems. While they addressed questions of language and translation in their writing, their research by no means supported the idea of automatic translation (though the correspondence between Mersenne and Descartes regularly mentioned the topic of translation). Leibniz's and Descartes' work, as well as the coding systems that followed them, were sources of inspiration for various researchers (and are often cited in the writings of the pioneers of machine translation), but they do not appear to have ever been used for the development of real systems.

Artificial Languages

The notion of universal language brings to mind artificial languages, of which Volapuk and Esperanto are the most well known. Volapuk is an artificial language invented in 1879 by Johann Martin Schleyer (1831–1912), while Esperanto was invented by Ludwik Lejzer Zamenhof (1859–1917) with the goal of facilitating communication between people with different mother tongues. Zamenhof published his project, called Lingvo Internacia (International Language), in 1887, under the pseudonym of Doktoro Esperanto ("Doctor who hopes"), the name by which the language became popular afterwards. All these projects emerged at the end of the nineteenth century in order to facilitate trade and peaceful cooperation between populations.

Artificial languages remain as a source of inspiration more than a real resource actually used in automatic translation systems.

Although these projects resulted in relatively advanced proposals with vocabularies and grammar systems, they have rarely been actively used for machine translation. Esperanto was used during the 1980s in the European Distributed Translation Language project and within the Fujitsu company in Japan, but these two projects were not completed. Artificial languages thus remain as a source of inspiration more than a resource actually used in automatic translation systems. One reason is probably that Esperanto remains a language designed for humans (Esperanto being itself based on various existing European languages): it does not have the characteristics of a language intended to be directly manipulated by computers. During the 1990s, the Universal Networking Language project aimed to develop such an artificial language to be used directly by computers, but it also remains to date relatively little used.

The Development of Mechanical Translation Systems between the Two World Wars

During the 1930s, two researchers devised mechanical systems oriented toward multilingual dictionaries and semiautomatic translation (for more information, see Hutchins, 2004).

Artsrouni's Mechanical Brain

The first attempt was the work of Georges Artsrouni, a French engineer of Armenian origin who had completed his studies in Russia before emigrating to France in 1922. In July 1933, he filed a patent application for a "mechanical brain": it was not so much a predecessor of modern computers as a machine to store and retrieve various types of information automatically. Two prototypes were built (probably between 1932 and 1935) and aroused great interest during public demonstrations. The machine even received a "grand prix" at the 1937 Universal Exposition in Paris (another prototype was built but never completed; the two existing models are stored at the Musée des Arts et Métiers in Paris).

In the late 1930s, various organizations handling large amounts of information showed considerable interest in this machine (in his patent, Artsrouni mentions that his machine could automate the consultation of railway schedules, telephone directories, and the search for words in dictionaries). Nevertheless, World War II prevented these contracts from succeeding. Finally, the emergence of computers after the war made these purely mechanical machines obsolete.

Artsrouni's system was not specifically dedicated to translation, though the inventor stressed from the beginning that this field was one of the most promising. The machine could store linguistic data (i.e., simple words) in different languages on a simple strip of paper. Each word was encoded in a unique way thanks to a set of perforations along the paper strip according to the principles of punch cards. A keyboard was used to indicate to the machine the sought-after word, and it could then automatically find the corresponding translations from the coding strip.

The system did not allow Artsrouni to go any further. He was not a linguist and never addressed the difficulties of machine translation, but his archives clearly demonstrated that he was one of the first to invent a completely automatic system based on multilingual dictionaries. He also thought of fairly realistic uses for his machine; for example, telegrams written in an elliptical style that would fit well with a word-for-word translation. Artsrouni also planned to directly store more complex linguistic units such as compound words for his machine: the only limit was the time and effort needed to encode the data.

Smirnov-Trokanskij's Assisted Translation Environment
Petr Petrovitch Smirnov-Trojanskij (1894–1950), who worked as a professor in Russia, filed a patent for a machine that would select and code words for translation between several languages. The machine was probably never developed as a prototype.

Smirnov-Trojanskij invented a workspace that to a certain extent was close to Artsrouni's machine: a mechanism specified a word to the machine, which was then capable of presenting translations for various available languages. Smirnov-Trojanskij's invention was only concerned with translation, unlike the machine developed by Artsrouni.

What makes Smirnov-Trojanskij's invention remarkable is that it goes beyond the simple coding of words and their translation. He imagined a system of 200 primitives capable of representing the function of a word in a sentence, in order to generate the correct translation in the target language (Smirnov-Trojanskij was interested in Russian, where nouns and adjectives are inflected to reflect their function in the sentence). The analyst had to specify whether the word to be translated was the subject or the object, whether the verb was in the present or imperfect tense, and so on. The machine then took over, selecting the correct word form for the translation.

The invention focused on a workspace, rather than on a simple device: Smirnov-Trojanskij's system was designed in such a way that a translator could first simply look for translation elements at word level with the help of the device. A professional text editor or a translator then intervened at the very end to edit the text and make corrections from a stylistic point of view. The difficulties of machine translation are not described in detail in his proposal, but this project is interesting in that Trokanskij envisioned an environment for assisted translation rather than a completely automatic process. We will see in the following chapters that the quality of automatically obtained translations remains a major issue, along with the way in which machine translations could be efficiently corrected by human editors.

It should be noted that, despite the considerable interest of their proposals, these two inventors have remained largely ignored. Artsrouni's system was not continued after the war, as it was clear that the future would lie with electronic machines (much more powerful than mechanical machines). Smirnov-Trojanskij's work environment, which never produced an operating system, was also largely ignored in favor of completely automatic translating systems.

THE BEGINNINGS OF MACHINE TRANSLATION: THE FIRST RULE-BASED SYSTEMS

The postwar period saw the advent of the first computers, and machine translation was immediately considered a key application. Several factors explain this keen interest: first and foremost, a pressing need (i.e., the need to automatically translate texts from foreign sources in the context of the Cold War), and secondly, strong theoretical issues (i.e., the question of how language works). Furthermore, progress in the field of cryptology during the war gave a glimpse of a possible solution: couldn't a document in a foreign language be considered an encrypted document that needed to be translated into an intelligible language? However, the first practitioners in the field quickly faced the limitations of the first computers. As a result, they developed a pragmatic approach based on bilingual dictionaries and transfer rules, making it possible to change word order according to the specificities of the target language.

These systems can include thousands of rules and are thus highly sophisticated, but are then hard to maintain. This approach, known as the rule-based approach, has been the dominant one for decades, and it continues to be popular today.

The Precursors

The first research attempts in the domain of machine translation were made in the United Kingdom, where Andrew Booth was concerned with data storage, and then in the United States with Warren Weaver, who sketched out a strategy for the domain with his seminal memorandum.

Early Experiments

Toward the end of the 1940s, Andrew Booth, from London University's Birkbeck College, became specifically interested in language processing by automatic means. His thinking was purely theoretical at the outset, since the first computers were being developed at the same time. The laboratory at Birkbeck College was an important research center on data storage and access. The size of electronic dictionaries was to cause major issues for decades due to the small storage capacity available on early computers. Booth also furthered research concerning machine translation and voice recognition.

In order to limit the number of entries within a dictionary (as in a standard dictionary, for example, where only the infinitive of verbs is recorded rather than all their inflected forms), Booth also took an interest in morphology. His algorithm searched only for sequences of characters: if a word was unknown—that is, if it was not included as such in the dictionary—the system tried to gradually remove letters from the end of the word in order to eventually find a known word form (for example, "*run*" from "*running*"). Despite its apparent simplicity, this technique works relatively well for English and continues to be used, particularly by search engines. The technique, called "stemming," makes it possible to get pseudo-roots for words without having to perform an advanced morphological analysis. Martin Porter popularized this technique in 1980 for search engines, and the technique is thus now known as "the Porter stemming algorithm."

This research was, to some extent, a continuation of Artsrouni's and Trojanskij's work on how to store multilingual dictionaries using mechanical means. Booth improved on these early attempts by adding a search algorithm that foreshadowed research on dictionary storage and management. With Richard H. Richens, he also created a word-for-word translation system based on bilingual dictionaries. These propositions were the first step toward a global approach to automatic translation but were quickly recognized as too simplistic, particularly by Weaver.

Weaver's Memorandum
The father of machine translation—and more generally of natural language processing—is unquestionably Warren Weaver. Along with Claude Shannon, he was the author of a mathematical model of communication in 1949. His proposal was very general and therefore applicable to many contexts.

In Weaver and Shannon's model, a message is first encoded by a source (which can be a human or a machine), sent, and then decoded by a receiver. For example, a message can be coded in Morse code, transmitted by radio, and then decoded in order to be comprehensible by a human. This model is the foundation of cryptography (encoding, transmission, and then decoding of the message) but can also be applied to communication in general: an idea, in order to be shared, must be "encoded," that is, "put into words," and transmitted to a hearer, who must then "decode" the message in order to understand its meaning. The same goes for translation, which can be seen as decoding a given text (the text is considered "encoded" in an unknown language: in order to be comprehensible, it must therefore be translated; in other words, decoded in the target language).

Beginning in 1947, Weaver corresponded with the cyberneticist Norbert Wiener concerning machine translation. He proposed that translation could be considered a "decoding" problem:

> One naturally wonders if the problem of translation could conceivably be treated as a problem in cryptography. When I look at an article in Russian, I say: "This is really written in English, but it has been coded in some strange symbols. I will now proceed to decode."[1]

For Wiener, the automation of translation was not directly possible, because language is made up of a large number of words that are either too vague or too ambiguous (in other words, one cannot translate by assuming simple and direct equivalences at word level). He wrote:

> As to the problem of mechanical translation, I frankly am afraid that the boundaries of words in different languages are too vague ... to make any quasi-mechanical translation scheme very hopeful.[2]

Wiener's notion of "boundaries of words" refers to the fact that a word like "*avocat*" in French has at least two meanings and thus at least two possible translations in English: "*avocado*" or "*lawyer*." This scenario, which is far from exceptional, is in fact omnipresent in language, since the majority of words have several meanings, and since the meaning of words is different in each given language (moreover, we can observe that any time the word "*avocat*" refers to a man who practices law, it refers to a "*lawyer*" but that, on the contrary, "*lawyer*" does not always refer to an "*avocat*": the word can also refer to other types of magistrates!). As a result, determining the meaning of a word and its possible translation in a given language seemed to be an almost insurmountable problem for Wiener, as it involved handling tens of thousands of "word meaning" pairs as well as actively determining the meaning of each word in context, at a time when computers still had very limited computational power and memory capacity.

Despite Wiener's doubts, Weaver carried on with his idea, and in 1949 he drafted a brief text expressing his thoughts on the subject. He specifically mentioned that words are often ambiguous, that their meaning depends on context, and that word-for-word translation is not a sufficient basis for high-quality results (Weaver was also corresponding with Booth about his research, and as a result became aware of the limitations of word-for-word translation). Weaver's reservations were not completely ignored but were largely discounted, which would have consequences in the future.

Weaver's text, entitled "Translation," is generally considered the starting point of research in this field. The memorandum was very influential, because in it Weaver developed ideas that were highly innovative for the time, but also because he was closely involved with an organization that financed research.[3] His influence was as much scientific as it was political.

Weaver proposed four specific principles in order to avoid the basic errors of a word-for-word translation:

1. Analyzing the context of words should make it possible to determine their precise meaning. The size of the context to be taken into account should vary according to the nature of the word (Weaver claimed that only a few nouns, verbs, and adjectives need to be disambiguated), but also possibly according to the topic and the genre of the text to be translated, if these elements are known.

2. It should be possible to determine a set of logical and recursive rules to solve the problem of machine translation, he wrote, "insofar as written language is an expression of logical character." According to Weaver, this excludes "alogical elements in language" such as "intuitive sense of style, emotional content, etc.," but machine translation can nevertheless be considered for the most part as a logical deduction problem.

3. Shannon's model of communication could probably provide useful methods for machine translation, since it had already proven useful "for solving almost any

cryptographic problem." In Weaver's words: "It is very tempting to say that a book written in Chinese is simply a book written in English which was coded into the Chinese code."

4. Languages can be described with universal elements that may help facilitate the translation process. Rather than directly translating from Chinese to Arabic or from Russian to Portuguese, it is probably best to search for a more universal and abstract representation that avoids any errors due to a verbatim rendering or to ambiguity.[4]

Each of these points deserves a closer look, as Weaver's suggestions are for the most part still being explored today. The first principle highlights the fact that most ambiguities can be solved by looking at the near context, which is the approach still used today. This is not enough to solve all kinds of ambiguities, but it is enough to solve most of them. However, the memorandum underestimated the problem of ambiguity. Weaver wrote: "Ambiguity, moreover, attaches primarily to nouns, verbs, and adjectives; and actually (at least so I suppose) to relatively few nouns, verbs, and adjectives." We now know that ambiguity is the most pervasive problem in natural language processing and applies to nearly all kinds of words, which makes ambiguity a much bigger problem than initially thought.

Ambiguity is the most pervasive problem in natural language processing and applies to nearly all kinds of words, which makes ambiguity a much bigger problem than initially thought.

The second principle was based on work done in logic and had a profound influence on the concept of formal grammar, which is used for analyzing artificial languages (particularly programming languages) as well as natural languages.

The third principle focuses on the comparison with cryptography, which at the time was a very popular research area due to the war. It highlights the statistical nature of language and the fact that computers could help solve difficult problems, especially in semantics. The following decades saw the development of logical approaches in language processing, and statistics were generally assumed to be too crude or even useless for the problem. The revival of statistical approaches in natural language processing in the 1990s showed how right Weaver was, but this kind of approach requires large amounts of data, which explains why this proposal did not become popular before then.

Finally, the last principle inspired numerous research projects aiming at developing interlingual representations, addressing the semantic content of sentences, and disregarding the particularities of each language.

Weaver mentioned several times in the memorandum that his point of view reflected his personal thoughts, which were not those of a linguist ("I have worried a good deal about the probable naïveté of the ideas here presented"). For him they were food for thought, most likely naïve, which should be reviewed by experts on the subject. Yet the memorandum was in fact very far-sighted, and that is what has ensured its remarkable posterity. It highlighted ideas that were explored during decades to come by symbolic approaches (i.e., the need for accurate semantic representations or for formal rules) as well as statistical ones (i.e., the fact that statistics are more powerful than symbolic rules to resolve ambiguities).

The implementation of the proposed techniques, however, required efforts that went beyond anything the pioneers of machine translation had ever imagined. In particular, the inherent ambiguity of natural languages showed that traditional encryption models were not sufficient to render the complexity of automatic translation.

The Real Beginnings of Machine Translation (1950–1960)

Weaver's memorandum and the perspectives it opened, as well as the proximity of the author to funding agencies, were the driving forces in the rapid development of research within this domain.

The Early Days
In the early 1950s, several researchers started to become interested in machine translation, which seemed to be

both a useful and logical application at the time. As already mentioned, two elements in particular played a determining role: (i) the work done on cryptography seemed then, following Weaver's ideas, to form a solid foundation for machine translation seen as a coding and decoding problem; (ii) the context of the Cold War also contributed to emphasizing the need for translation, especially from Russian into English in the Western world (and vice versa in the Soviet world).

It was in this context that an Israeli researcher, Joshua Bar-Hillel, played a leading role in the development of machine translation in the United States during the 1950s. Bar-Hillel spent two years at MIT in 1951–1953, working as a post-doctorate fellow under Rudolf Carnap. Bar-Hillel had actually first corresponded with Carnap while he was working on his thesis in Israel back in the 1940s. Carnap, the German philosopher who later became a naturalized American, had developed a "logical syntax of language," which seemed to pave the road toward a logical formalization of natural languages.

Bar-Hillel then naturally became interested in machine translation. He quickly became a major figure in the field and benefited from grants that allowed him to visit major laboratories in the United States (research teams were being formed at the time and were relatively scattered among various American universities). Upon his return to MIT, Bar-Hillel drafted a document pointing out the interest of the field but also highlighting the difficulties of its task (this document in some ways echoed the conversation between Wiener and Weaver that had occurred only a few years earlier). Immediately afterwards, he organized the field's first conference at MIT in June 1952.

The majority of researchers active in the field attended the conference at MIT. The attendees were clearly excited and emphasized the need to attract large amounts of funding, given that machine translation required human capacities, and especially access to computers that were extremely expensive at the time. In order to promote machine translation, the representative from Georgetown University (a major research center and pioneer in the field) suggested that a demonstration be organized as soon as possible in order to show the feasibility of the project and attract funding.

In 1954, the research team at Georgetown University, along with IBM, led the first demonstration in support of machine translation based on a system developed jointly by the two teams. A set of 49 Russian sentences was translated into English using a relatively simple dictionary (a dictionary of only 250 words and six grammar rules). The impact of the demonstration was considerable and contributed to the increase in financial support for machine translation. There was also extensive media coverage of the event, which helped attract public attention.

American funding agencies gradually began to support a number of groups working on machine translation, primarily in the United States and the United Kingdom. The 1954 demonstration also grabbed the attention of the U.S.S.R. and several Soviet research teams, who became involved in the field from 1955 on. The field of machine translation was institutionalized with regular conferences and a specialized journal, *Mechanical Translation*, first issued in 1954.

The Development of the First Rule-Based Systems: Turmoil and Enthusiasm

The majority of research teams at the time had very limited access to computers, which were not widespread, especially in the U.S.S.R. In fact, most of the work remained theoretical and offered approaches that could "mechanize" the translation process, without ever being put into practice.

Schematically, it can be said that two lines of research were pursued: in the first, the "pragmatic" route aimed at quickly producing results, even if those results were not perfect. The systems were essentially based upon a direct translation approach: bilingual dictionaries first provided a verbatim translation, and then reordering rules were applied to accommodate the word order of the target language. In other words, a dictionary was first used to find equivalences between words and then basic reordering rules were used to control certain phenomena, such as noun-adjective phrases in French that must be translated as adjective-noun in English (*"voiture rouge"* → *"red car"*).

At the same time, those in favor of a more theoretical approach highlighted the limits of the direct approach. Numerous proposals were then made to promote an analysis of the source text before the translation process, and to develop transfer rules operating at the syntactic or semantic level (and not just at word level). In this regard, it is relevant that the notion of formal grammar became prominent in the 1950s, mainly with work by Noam Chomsky. Certain research centers were also interested in the idea of a pivot language (i.e., an approach in which a particular language is used as a kind of intermediate representation between the source language and a target language), or even in the idea of an interlingua (i.e., an artificial language that offers an abstract representation of the sentences to be translated). In both cases, the approach consisted in encoding all the necessary information needed for translation in a specific representation model. The interlingua is thus an artificial language that has nothing to do with any existing language, whereas a pivot language uses an existing language (generally English) for this representation.

Several research groups (in Washington and at Harvard and the Rand Corporation, for example) made every effort to develop large bilingual dictionaries (Russian-English), either manually or with the help of a statistical analysis of specific corpora, which helped ensure that the most frequent or the most important words would be processed first. Polysemy—the fact that a single word can have several meanings, such as "*bank*," which can refer to a financial institution or the side of a river[5]—was seen from the beginning as one of the major problems to solve. The simplest approach would be to include only the most expected word meanings in the dictionary. While doing so would solve the problem, clearly it is too extreme: the results of the direct approach (with no semantic disambiguation process) are therefore unsatisfactory. The translation fragments provided, even if highly imperfect, can nonetheless be useful if the reader has no knowledge of the target language, or can serve as a basic "translation memory" by providing regular equivalences between languages.[6]

In order to solve the ambiguity issue, many research teams gradually enriched the content of their electronic dictionaries. For example, the University of Washington added contextual information to words so that ambiguities could be resolved without a full syntactic analysis. Vocabulary was also partitioned by domain (the idea being that "*bank*" will probably not have the same meaning in a financial corpus as in an environment corpus, or at least that significant statistical differences will help in the disambiguation task) and multiword expressions were gradually added (to avoid some sources of ambiguity with simple words). The approach may sometimes seem ad hoc, but it should be noted that current techniques still have significant similarities with the strategies identified in the 1950s: a local analysis is often enough to determine the category of a word, even its meaning. Storing multiword expressions and taking into account the domain does indeed help to drastically reduce the ambiguity problem: "one sense per discourse" even became a popular slogan in the field during the 1990s.

At the same time, as already seen, more fundamental work on parsing—that is, the automatic syntactic analysis of sentences—began to appear. Chomsky independently developed his own work on syntax but did not have any significant influence on machine translation until the 1960s. However, the need for a formal analysis of source languages gradually became a mainstream idea. Several research groups actively developed this strategy for machine translation in the late 1950s. One must keep in mind that, at the time, the formal analysis of languages targeted both programming (i.e., artificial) and natural languages. It was therefore not self-evident that, since natural language processing ultimately has very little to do with programming languages, the ambiguity issue in natural languages meant that very specific strategies had to be designed for this field. The first specific formalisms were then defined: they were "stratified" (using an expression coined by Sydney Lamb, who defined a "stratificational grammar"), ranging from low-level information (word categories, morphosyntactic features) to high-level information (the meaning of words and their possible context). These pioneering studies were quite valuable and instructive. A number of research groups recognized the great difficulty of the task, especially the "semantic barrier" that Wiener, and particularly Bar-Hillel, had anticipated since the beginning of the 1950s (see above).

Beyond the United States

Before continuing to the next period, we must point out the research done outside the United States. Since the mid-1950s, Cambridge University, through the Cambridge Language Research Unit, had benefited from American subsidies and developed one of the first interlingual systems, called NUDE. According to its designer, Richard Richens, NUDE was a "notational interlingua ... constructed so as to represent the ideas of any base [source language] passage divested of all lexical and syntactical peculiarities; for which reason it [was] called Nude" (Richens, 1956, cited in Sparck Jones, 2000). The NUDE interlingua aimed to define each word by means of a set of universal primitives (core meanings that can be assembled to represent the meaning of complex ideas expressed in various ways, depending on the natural language in question). The implementation of this approach remained limited and seems to have suffered from a poor link to syntax (so that it was not clear how a NUDE representation could be derived from an actual text). Nonetheless, this proposal remained important since it opened a new strand of research and popularized the idea of universal semantic primitives that can be found in numerous linguistic theories all over the world. More generally, the Cambridge group prioritized the development of semantic resources (word lattices) and techniques that were partially rediscovered years later for semantic disambiguation (i.e., choosing the meaning of ambiguous words according to the context). Of course, it did not develop definite answers to questions that are still largely debated today, but it was a pioneering and influential group in the study of semantics during a time when attention was primarily focused on syntax.

Other research groups in the field of machine translation appeared toward the end of the 1950s, for example in 1956 in Japan and 1957 in China. In France, the interest was clear from the late 1950s on, and two centers were then created by the French National Center for Scientific Research (CNRS), in Paris and Grenoble. The interest in machine translation was simultaneous with the first computers intended for university centers in France, and was, therefore, the real beginning of computer science in the country. The two centers were called Centre d'Études sur la Traduction Automatique, or CETA: CETAP was located in Paris, and CETAG in Grenoble. The Parisian center encountered financial problems from very early on and had to bear the consequences of the criticism of machine translation that was emerging in the United States. In fact, the center closed a few years later and some researchers, such as Maurice Gross, turned to computational linguistics, stressing the need to first develop rich linguistic resources that offer a broad and systematic description of language. The Grenoble center has survived to the present day and developed an original interlingual approach. As a result, Bernard Vauquois, who led the CETAG center in Grenoble and proposed several influential ideas in the field, became one of the major figures until his death in 1985, although machine translation was then no longer as popular elsewhere in the world.

Finally, a few words must be said about the research being carried out during the same period in the USSR. The Georgetown-IBM demonstration in 1954 made a strong impression in the Soviet world, which immediately decided to launch research in this domain. Several groups rapidly began working on problems in machine translation, primarily in Moscow but also in Leningrad and in other "sister countries." The first congress on machine translation organized in Moscow in 1958 was attended by about 340 participants from 79 different institutions. The approaches were as diverse as in the United States, but the majority of the research remained theoretical due to the unavailability of computers. The few groups lucky enough to have access to computers essentially developed empirical and direct approaches based on bilingual dictionaries.

At the same time, many theoretical studies specified strategies for an automatic syntactic analysis, but also for the coding of semantic information. Linguistic theories dating from this period still have a large audience to this day. The work of Igor Mel'čuk and Yuri Apresjan, in particular, is well known today, including outside of the former Soviet world, especially because Mel'čuk settled in Canada in the late 1970s.

A Period of Disenchantment (1960–1964)

The end of the 1950s saw the first doubts expressed about the feasibility and even possibility of obtaining correct translations as the outcome of an automated process.

Bar-Hillel's Criticism

Bar-Hillel, who had returned from Israel at the end of his post-doctorate position in 1953, had the opportunity to return to the United States a few years later for a new research residency (1958–1960). In September 1958, during his trip to the United States, he presented a paper to the University of Namur entitled "Some linguistic obstacles to machine translation." In this text, Bar-Hillel lists some linguistic issues that he considered to be fundamental problems for machine translation, since no system was then able to solve them. In his opinion, the models at the time were too simple and needed to be replaced by models that would better account for the structure of the sentences to be analyzed.[7] Furthermore, according to Bar-Hillel, the transfer rules required to translate between genetically distant languages had to be complex and required formalisms yet to be invented. After his conference in Namur, Bar-Hillel continued his trip to the United States to assess the research being conducted in the field.

There, he drafted the famous technical report entitled "Report on the State of Machine Translation in the United States and Great Britain" (February 1959) on behalf of the U.S. Office of Naval Research. The report delivered an extremely negative assessment of the ongoing work, without considering the very limited history of the field (most of the groups had only existed for a few years). All the research groups were listed by name and severely criticized.

Practically, Bar-Hillel noted that, on the one hand, translation needs a complete syntactic analysis of a text, which was not completely obvious for all the groups involved in the field at the time. On the other hand, translation needs to resolve semantic ambiguities, which was beyond the state of the art at the time and did not seem solvable in the medium term. An appendix of the report had an evocative title ("A demonstration of the non-feasibility of fully automatic, high-quality translation," see Bar-Hillel, 1958 and 1959) and was intended to show that the meaning of some ambiguous words cannot be determined, even when taking into account the context, which suffices to invalidate the goal of high-quality machine translation. Bar-Hillel used the following well-known example:

> "Little John was looking for his toy box. Finally, he found it. The box was in the pen. John was very happy."

In order to understand the sentence, one must realize that the word "*pen*" refers to a small enclosure in which a child plays, and by no means to a writing utensil. Yet there is nothing in this context that enables the reader to infer this meaning for "*pen*," which is much less common than an implement for writing. According to Bar-Hillel, such an example demonstrated the impossibility for any system to solve this kind of problem, which he believed would happen quite frequently. It was therefore impossible to envision a completely automatic, high-quality translation in the short or medium term (FAHQT, or fully automated high-quality translation; also found as FAHQMT, or fully automated high-quality machine translation).

Instead of automatic translation, Bar-Hillel recommended that researchers turn toward computer-assisted translation systems, which constitute a relatively different project, clearly less exciting scientifically speaking than the idea of an entirely automatic system. Bar-Hillel called for the development of translation aids, which would significantly help the productivity of translators by proposing suitable and efficient tools, specifically for the pre- and post-edition stages (preparing the text for translation; correcting translation errors). Since the goal is then to help translators, system outputs must be relatively different from those of traditional machine translation systems: for example, it is generally better to present the translator with suggestions of translations rather than directly produce a text, which would be difficult to correct.

Discussion

As we have seen, the 1950s, which had gotten off to a flying start, subsequently ended on the first doubts regarding the feasibility of machine translation.

Bar-Hillel's report focused on real problems that had been underestimated up until that point. The approaches considered initially failed largely due to oversimplification: the hopes of advancing rapidly were too optimistic, and initial results proved disappointing. The 1954 demonstration was based on sentences that were prepared in advance, with a familiar vocabulary and limited ambiguity that clearly had little to do with the reality of the task, which concerned previously unseen texts from any domain. Similarly, most research groups in the 1950s did not realize the need for a syntactic or semantic analysis, and therefore did not evaluate the difficulty of the task properly. Finally, the idea of a translation aid was more realistic if the goal was to provide a quick operational response, but this had little to do with the advancement of machine translation.

The research in the 1950s nonetheless established the field of machine translation. The setbacks, or at least the limitations, of these early systems revealed the complexity of natural language processing. In some ways, they sparked off numerous research projects pursued in the following decades. Although machine translation was too ambitious at the time, the research was not useless. We must also keep in mind the relative lack of computers and their limited capabilities—at the time of punch cards—which drastically restricted possibilities for experimentation.

However, Bar-Hillel's report raised doubts not only for those funding the research, but also for researchers themselves. Several leading figures left the field at the beginning of the 1960s and moved to research in linguistics, computer science, or information theory. Certain researchers were even more negative than Bar-Hillel himself on machine translation.

Alternatively, the demonstration gave a glimpse of the numerous problems the first projects had underestimated. Georgetown's and IBM's attempt to industrialize practical solutions yielded very poor results.

All of this brought funding agencies in 1964 to ask an independent committee for an evaluation report. Their request led to the famous ALPAC report, published in 1966.

THE 1966 ALPAC REPORT AND ITS CONSEQUENCES

Published in November 1966, the ALPAC Report was a milestone in the history of machine translation: its influence was significant, but is now perhaps a bit overestimated. At the beginning of 1964, the funding agencies that had been financing machine translation programs in the United States[1] commissioned a group of experts to create the report. Now known for having highlighted the failures of work conducted since the late 1940s, the report was clearly a follow-up to Bar-Hillel's observations.

The ALPAC Report can easily be found online,[2] and there have been several articles on the history and impact of the report (among others, see Hutchins, "ALPAC: The (In)famous Report," 2003). Here we will discuss the content of the report and the research conducted up to the end of the 1980s, in the years following its publication.

Content of the Report

The title of the report itself is "Languages and Machines: Computers in Translation and Linguistics." The key focus of this short report was in fact translation needs: the usefulness of translation for relevant agencies—mostly the public sector and businesses related to security and defense; the report observes that the majority of requested translations are of negligible interest, and ultimately are either partially read or not read at all—and the costs associated with these translations. The discussion on machine translations takes up only a short five-page chapter.

The Automatic Language Processing Advisory Committee (ALPAC) was directed by John R. Pierce, an information and communication theory specialist (he had worked with Claude Shannon in particular; see chapter 5). In addition to Pierce, the committee was made up of linguists, artificial intelligence specialists, and a psychologist. None of the committee members were working on machine translation at the time of the report, though two of the members (David G. Hays and Anthony G. Oettinger) had previously been active in the field. The committee did, however, interview several machine translation specialists (Paul Garvin, Jules Mersel, and Gilbert King as representatives from private companies working on machine translation, as well as Winfred P. Lehmann from the University of Texas).

The introduction of the report mentioned two reasons that could justify financing the research out of public funds (with the exception of the National Science Foundation, the agencies financing the research in the United States were closely related to defense and intelligence agencies). Those two reasons were as follows:

(i)   if it served as a long-term, fundamental research program that would have a significant impact ("research in an intellectually challenging field that is broadly relevant to the mission of the supporting agency");

(ii) or, on the contrary, if the research aimed to solve a practical problem ("research and development with a clear promise of effecting early cost reductions, or substantially improving performance, or meeting an operational need").

The report indicated that research on machine translation clearly corresponded to the second reason (obtaining quick and effective methods at a lower cost in a relatively short period of time) and therefore proposed an evaluation of the field in this regard. This was obviously a major bias, in that failing to develop a practical and efficient solution within a short period of time did not demonstrate the uselessness of the research being conducted. Ultimately, the nature of the agencies funding the research constituted a major bias for the evaluation process.

The research teams also suffered the consequences of failing to fulfill the promises they had made since the beginning of research in the field. The 1954 demonstration (see chapter 5) suggested that a practical solution was within reach. Yet industrial attempts and public demonstrations at the end of the 1950s and from the beginning of the 1960s showed that they were far from finding a solution. In fact, this contradicts the discourse from several years earlier, when the groups suggested that machine translation could yield operational results within a few months.

We must therefore keep in mind that the report was above all geared toward evaluating the possibility of obtaining high-quality machine translation in the near future (FAHQMT, or fully automatic high-quality machine translation; see chapter 5). This gave a particular twist to the report, and later had a significant impact on the field. This perspective also explains why the first half of the report examined the large quantity of translations ordered by the agencies involved, the number of available translators, and the costs incurred. Upon reading the report, it is very clear that a practical issue was under evaluation, and that the main yardstick was cost! Research perspectives were the least of the report authors' concerns.

In fact, the report concluded, in terms of costs, a human translator was more affordable than machine translation. At the time, human translators allowed for better and faster translations, as there was no need for additional editing (correcting a text translated entirely by machine often took longer than a direct translation carried out by an experienced translator). The report only considered translations from Russian to English, and, as a result, concluded that the need for Russian to English translation was limited. The largest "consumers" of Russian translation would do better to learn the language itself, the authors suggested. Incidentally, the report seemed overly optimistic, given that it suggested a few weeks was enough to acquire a good command of a foreign language![3]

The report clearly shows that, in the mid-1960s, there was no need for machine translation. According to the report, this field had no practical interest given that there were no appropriate systems to carry out the task. The original text put it very bluntly: "There is no emergency in the field of translation. The problem is not to meet some nonexistent need through nonexistent machine translation."

The report then addressed the more general question of funding machine translation. The report began with a fairly standard definition: machine translation "presumably means going by algorithm from machine-readable source text to useful target text, without recourse to human translation or editing." The report immediately concluded that no type of automated system existed at the time of drafting the report and that no such system was

"There is no emergency in the field of translation. The problem is not to meet some nonexistent need through nonexistent machine translation." [Alpac Report, 1966] conceivable in the near future.[4] Georgetown's system was specifically mentioned: after eight years of funding, the system was still unable to produce a proper translation. A professional translator still had to step in and correct the translation errors. The report emphasized that while machine translations most commonly produced a decipherable text, they were equally likely to contain mistranslations and errors. The more faults a translation contains, the more difficult it becomes to manipulate and correct the text.

To illustrate the point, the report included four translation results from Russian to English using four of the era's machine translation systems. The translations were mediocre at best.

Direct Consequences of the Report

In his 1996 article, Hutchins recalled the notoriety of the ALPAC report, pointing out that its importance had probably been exaggerated. Research funding had already decreased at the beginning of the 1960s, a situation for which Bar-Hillel's 1959 report was partially responsible. Consequently, the number of groups working in the field of machine translation in 1966 was much lower compared to 10 years earlier (Washington University and Michigan University, as well as Harvard, had stopped their research projects in 1962; Georgetown University, specifically mentioned in the report, had not received any financial support since 1961). Other projects were pursued after 1966, at Wayne State University and the University of Texas in particular (up until the 1970s in both cases). The report simply confirmed the decision to drastically cut back on financial support for the field of machine translation.

Hutchins also emphasized the bias of the report: it only took into account translations from Russian to English executed by American agencies, and it ignored the problems of multilingualism beyond this particular context. The nature of the report, in addition to its ambitions, needs to be examined as a whole. It was very clear that the automatic translation systems of the mid-1960s were not capable of directly solving industrial needs. Nevertheless, machine translation drew attention to many scientific issues that were hardly mentioned in the report. The report even amplified Bar-Hillel's conclusions that a completely automatic translation system was not possible in the near future.

On a positive note, the ALPAC report did express interest in computer-assisted translations, an idea Bar-Hillel also supported. The report also pointed out, rather indirectly, the need for more fundamental research on the automatic analysis of languages. It should be noted, for example, that even Hays and Oettinger, members of the ALPAC committee, had stopped their research on automatic translations a few years earlier and instead focused on syntax and parsing. Thus, Oettinger's 1963 report, entitled "The State of the Art of Automatic Language Translation: An Appraisal," broadly recapped Bar-Hillel's conclusions concerning automatic translation, but also revealed a clear interest for natural language processing.[5]

1965–1990: A Long Pause

The period following the publication of the ALPAC report represented a break from research in the English-speaking world. Other countries continued to finance research teams, while the first commercial systems began to emerge. The technical innovations during this period were limited, following their abundance in the first decade.

A More Widespread Research Effort

The ALPAC report cemented the lack of funding in the United States in the field of automatic translation during the mid-1960s. In the United States, two groups nonetheless continued research on automatic translation (at the aforementioned Wayne State University and the University of Texas), but even there, the emphasis was on syntactic analysis to make it possible to develop rich transfer rules between languages. Other groups (like Oettinger's group at Harvard, for example) completely abandoned automatic translation and turned toward syntactic analysis, which in some ways can be considered a logical continuation from the previous period.

Hutchins (2010) emphasizes that, contrary to the United States, numerous countries have to cope with a multilingual landscape, making it easier to justify the continuation of research in the field. Canada, in particular, opened a research center in Montreal in 1965, when the majority of American centers had already closed (in the 1970s, the center was called Traduction Automatique de l'Université de Montréal, or TAUM). The need to produce a large quantity of official documents in English and in French led to high costs, which created a strong incentive to launch research in the field. The group from Montreal quickly produced two important results: a formalism suitable for representing linguistic information, developed by Colmerauer (this formalism can be seen as a precursor of the Prolog programming language that has been since then very popular in computational linguistics and more generally in artificial intelligence) and, above all, probably the most well-known automatic translation system: TAUM-Météo (later referred to simply as Météo; see below).

In France, research continued in Grenoble, where the CETAG group (later known as CETA after the closing of the Parisian center; see chapter 5), under the direction of Vauquois during the 1960s, developed an original translation system in which syntactic relationships were represented in a language-independent logical formalism (though the system was not really interlingual, since it also used bilingual dictionaries). The research mainly focused on the translation of mathematical and physics texts from Russian to French. However, the system lacked flexibility: a problem at any level was enough to block the entire translation process. During the mid-1970s, Vauquois set off to develop a modular system, with the possibility of transferring linguistic information between two languages at different levels. This was to develop into the Ariane-78 system, and is reminiscent of the image of his triangle shown in chapter 3: an ideal translation would require a logical representation (i.e., the top of the triangle), but if an automatic translation system cannot reach this level of precision, a precise syntactic or semantic analysis is better than nothing.

In the same way that prevailing bilingualism drove Canada to finance a research center while the United States was turning away from machine translation, the need to produce translations between a growing number of languages within the European Union encouraged the European Commission to become interested in automatic translation in the 1970s. The European Union initially examined the first available commercial systems. That was how a company set up in the United States in 1968, Systran, came to present its system to the European Union in 1975. Systran then developed a prototype that integrated different languages from Europe, entering into a partnership agreement that continued throughout the 1980s. We will return to this topic when we examine Systran's history in greater detail in chapter 14. Also at the end of the 1970s, and largely under the leadership of Vauquois, a major European research program was launched: the Eurotra project. Active from 1978 until 1992, the project emphasized the syntactic level of analysis more than the development of bilingual dictionaries. The goals of the project—initially to create an operational system—were progressively scaled down and never gave birth to a successful system. It mainly resulted in several prototypes and in the emergence of new collaborations between European research institutes. Elsewhere in the world, particularly China, Japan, and the Soviet Union, several centers were created and carried out their own research during this period.

The eventual emergence of parallel corpora (i.e., pairs of translated texts) led to the invention of new methods for automatic translation, pushing this research area in multiple new directions. This will be the topic of discussion in the following chapters.

The First Commercial Systems

Some of the previously mentioned research groups produced prototypes that led to commercial or operational systems.

Montreal's research center developed the TAUM-Mé-téo system in the 1970s, which then became simply Météo and was run by John Chandioux, an independent developer from the TAUM group. Each day, the system translated the weather forecasts in Canada into two languages, French and English, on behalf of Environment Canada. The forecasts concerned not only the country but also each of the provinces that produced several daily forecasts, resulting in a significant volume of translations. The system was operational from 1977 until 2002, translating several hundreds of thousands of weather forecasts in total—about 30 million words a year during the 1990s. Although the system was relatively classic in design, it was the first to show the possibilities of obtaining operational solutions in restricted domains. The quality of the translated texts was good: very little post-editing work was required, thus allowing for reliable, robust, and regular translations. This system played an important role in promoting machine translation, especially during a period when the field was suffering from a rather tarnished reputation.

Through the 1970s and 1980s, other research groups established partnerships with manufacturers to develop specific translation solutions. For example, during the 1980s, the University of Texas teamed up with Siemens to develop Metal, a translation system initially aimed at the German-English language pair, then gradually adapted to other languages. In Japan, most companies in the software and hardware industry launched projects to produce operational systems between Japanese and English, but some also focused on other Asian languages, such as Chinese and Korean. The semiautomatic translation of short technical texts and product leaflets (i.e., the translation by means of an automatic translation system whose results can be revised by hand) was a primary commercial objective.

A final point to mention is the emergence of the first companies specifically dedicated to automatic translation since the late 1960s. First and foremost was Systran, founded in 1968 by Peter Toma, a former member of the group at Georgetown. Thanks to contracts with American defense organizations and its commercial partnership with the European Union, Systran quickly acquired a unique status within the field (for more information, see the history of Systran, told in chapter 14). Another example is the Logos Corporation, established in 1970 with the support of the American Ministry of Defense for the purpose of translating texts from English to Vietnamese. The context of the Vietnam War suddenly led to a period of increased need for translation into Vietnamese. Logos gradually expanded the number of languages processed over several decades until it became Systran's main competitor. The company closed in 2000, and only a translation program called OpenLogos remains; it is still available online as free software.

These companies demonstrated that there was a limited but real need for automatic translation. Translating texts (leaflets, manuals, etc.) into several languages is relatively complex and costly (it requires, for example, finding translators for different languages, making sure the translations remain up to date with product development, etc.). Small and medium-sized companies need to produce translations but cannot invest too much money in this. Hence, machine translation is often seen as a desirable technology from their point of view. Outside of this niche market, large public administrations and the defense and intelligence industries remained the primary clients of these companies. We will return to this topic at greater length when we take a look at the current machine translation market in chapter 14.

PARALLEL CORPORA AND SENTENCE ALIGNMENT

The 1980s saw an increase in the number of available electronic texts—texts directly accessible by computers. Among these texts, some were translations of each other and could therefore be "aligned," or matched at paragraph or sentence level. These aligned texts have always been an invaluable source of knowledge for human translators. It soon became clear, however, that automatic tools could also benefit from these new data. In fact, this increasingly large amount of data revived the field and, most importantly, completely revolutionized the approach used for machine translation.

The Notion of Parallel Corpora or Bi-texts

A parallel corpus is a corpus composed of a set of pairs of texts in a translation context. An aligned parallel pair of texts is called a *bi-text*, from bilingual text, while a *multi-text* includes multiple aligned translations.

This type of resource is popular among professional human translators. It is actually an extremely valuable source of knowledge: more than a bilingual dictionary, previous translations can provide examples of relevant translations according to the context. In the case of technical translation, the translator must generally use the terminology, the phraseology, and the style of previous translations for reasons of uniformity. It is therefore essential to have access to past translations. It is also important to know in which direction the translation was carried out (i.e., the source language) since the source is by definition the reference text.

Human translators thus generally have access to past translations through a tool called a "translation memory." A translation memory module makes it possible to store and retrieve fragments of translation from past work, generally through a powerful search engine. Past translations can be analyzed and tagged before being stored, which makes it possible to query the translation memory powerfully, rather than with basic keywords. There are several tools like this available on the market, primarily for professional translators.

Bi-texts were perceived very early on as an important source of knowledge for machine translation. Translation memories contain relevant translation fragments, since these tools store professional past translations. Beyond that, more and more bilingual texts are available on the Internet, so one can imagine the development of systems based uniquely on bilingual data from the Internet. Today, this is in fact the dominant approach in the field of machine translation.

Two types of approaches can be distinguished. On the one hand, the analysis of existing translations and their generalization according to various linguistic strategies can be used as a reservoir of knowledge for future translations. This is known as *example-based translation*, because in this approach previous translations are considered examples for new translations. On the other hand, with the increasing amount of translations available on the Internet, it is now possible to directly design statistical models for machine translation. This approach, known as *statistical machine translation*, is the most popular today.

Unlike a translation memory, which can be relatively small, automatic processing presumes the availability of an enormous amount of data. Robert Mercer, one of the pioneers of statistical translation,[1] proclaimed: "There is no data like more data." In other words, for Mercer as well as followers of the statistical approach, the best strategy for developing a system consists in accumulating as much data as possible. These data must be representative and diversified, but as these are qualitative criteria that are

"There is no data
like more data."
[Robert Mercer]

difficult to evaluate, it is the quantitative criterion that continues to prevail. In fact, it has been proven that the systems' performance regularly improves as more bi-texts are available to develop it.

Availability of Parallel Corpora

There are two major sources of bi-texts: on the one hand, corpora already available for two or more languages; the bi-texts may be aligned or not. On the other hand, for pairs of languages without adequate corpora, techniques have been developed to automatically develop such corpora, generally by collecting texts available on the web.

Existing Corpora

There are well-known sources of parallel texts. For example, the majority of countries and institutions that have several official languages have to produce official texts (legislative texts for example) in each of these different languages. This is generally a source of very valuable bilingual corpora, since the translation must be an accurate copy of the original. But since these texts are for the most part associated with legislative and legal fields, machine translation systems based on these data may not be very accurate for other domains or other genres.

The first experiments regarding the alignment of texts, in the 1980s, drew heavily on the Canadian Hansard, which records the official transcripts of Canadian parliamentary debates. The Canadian Hansard is aligned at text level and also at sentence level, which means that this corpus is an invaluable source of knowledge when translating between French and English (see figure 3).

French Text	English Text
J'ai fait cette comparaison et je tiens à m'arrêter sur ce point.	I have looked at this and I want to talk about it for a second.
L'article 11 du projet de loi crée tellement d'exceptions qu'il va bien au-delà de l'article 21 de la convention, au point de carrément compromettre l'objet même de celle-ci.	Clause 11 in the bill creates so many exceptions that it goes well beyond article 21 of the treaty and basically completely undercuts the intention of the convention itself.
Je cite l'article 21 de la convention.	I will read what article 21 says.
C'est assez simple:	It is pretty straightforward:
Chaque État partie encourage les États non parties à la présente Convention à la ratifier, l'accepter, l'approuver ou y adhérer [...]	Each State Party shall encourage States not party to this Convention to ratify, accept, approve or accede to this Convention [...]
Chaque État notifie aux gouvernements de tous les États non parties à la présente Convention.	Each State Party shall notify the governments of all States not party to this Convention.

Figure 3 An extract from the Hansard corpus aligned at sentence level.

Other corpora of the same type are available today, particularly corpora made of texts produced by European institutions. The European context is by nature highly multilingual and has already produced several invaluable resources, such as the Europarl corpus and the JRC-Acquis corpus. Both of these corpora include more than 20 languages from Europe. These corpora have been intensively used by machine translation systems and are easy to use since they are in fact already aligned at text and paragraph level, as well as occasionally at sentence level. They consist of several tens of millions of words for each language, but the size varies a great deal according to the language or the pair of languages in consideration (for example, Europarl contains 11 million words for Estonian, 33 million for Finnish, and 54 million words for both English and French).

Many other corpora exist, especially for other families of languages, although it is generally the "strongest" languages (meaning the most widely represented on the Internet) that are the most popular. However, these corpora are not always sufficient: most languages have very few or even no resources at all to develop a system. In this context, it is necessary to develop new corpora, and this is usually done on the web.

Automatic Creation of Parallel Corpora

Researchers early on sought to exploit the mass of texts available on the Internet to complete existing resources. The web is actually far more diverse than existing available parallel corpora, which are related to the legal domain for the most part, as already mentioned. The techniques for "harvesting" high-quality bilingual texts on the web are relatively simple. A harvesting system generally includes a "robot"—that is, a system capable of browsing the web by bouncing from page to page, while following the links mentioned on each webpage. Then, for each webpage, the system checks the language used and if an equivalent page in the target language exists.

The system begins with the rarer language of the two. For example, if the aim is to develop a bilingual corpus between Greek and English, it seems more appropriate to begin with websites written in Greek, which are fewer in number than websites in English. It should also be noted that few pages in English have a corresponding page in Greek, whereas the opposite situation is more likely; for example, websites of English universities rarely have a translation in Greek, whereas websites of Greek universities often have a translation in English.

For each website or page, two tricks can be used: first, the system searches for an equivalent at the website address (URL) level. For example, if one site corresponds with the URL http://my.website.com/gr/, the system will look for an equivalent such as http://my.website.com/en/—that is, a "mirror site" in the target language, identified by its URL. If this first strategy does not work, the system can search each webpage for a link toward the page in the target language, since multilingual sites often make it possible to navigate from one language to the other (these links are often identifiable through small icons featuring the country symbol of the target language). Once two websites have been identified as being translations of each other, the system has to control correspondences at the level of individual webpages. Several tools can be applied to check the language of the identified webpages if this was not previously done. Then, one can compare, for example, the length of the documents (if two documents or two texts are of very different sizes, they are probably not a reliable translation), the HTML structure (the two files must share the same structure), and so on.

These techniques have little to do with linguistics, but, when applied at web scale, they can allow for the extremely rapid development of large corpora in numerous languages from scratch. If the content of the selected webpages is closely monitored (for example, by starting with a list of specific URLs and then retrieving only those webpages that contain specific keywords), it is furthermore possible to obtain specialized corpora for different domains at a lower cost. Nonetheless, one must keep in mind that the process is entirely automatic: it does not guarantee the representativeness of the data nor the quality of the identified sources. In fact, nothing guarantees the quality of the bi-texts obtained in such a way. However, quantity goes together with quality: a given website may propose poor translations, but the consequences will be limited, since one can expect that a multitude of other websites will propose "good" translations, which means that bad translations will be statistically negligible and have no influence in the end. For the same reasons, a literary translation that is unique and original will also be discarded because it will not be statistically significant among all the other translation possibilities. This is not really a problem for machine translation, which looks for standard equivalents and does not attempt to produce originality.

Still, the limitations of this approach must be noted. Not all languages are well represented on the Internet, and this is especially true when searching for bilingual texts. In practice, in the majority of existing corpora, one of the languages is English, increasing the influence of this language. Despite the amount of available data, it is difficult to harvest enough data to develop a quality bilingual corpus, if one of the languages (target or source) is not English. I will return to this subject later in chapter 11.

Once the corpus has been built, it is necessary to align it at paragraph or sentence level, or both, for it to be usable by machine translation systems.

Sentence Alignment

In nearly all languages, a sentence is a linguistic unit that is syntactically and semantically autonomous (as opposed to a phrase or any other nonautonomous group of words). Consequently, natural language processing is often based on the notion of the sentence—particularly machine translation, which operates generally sentence by sentence, each being considered independently from the others.

Sentence alignment flourished toward the end of the 1980s and in the 1990s, during a time when more and more corpora were becoming available. Several kinds of applications using this type of resource were also beginning to appear: machine translation, of course, but also other multilingual applications, for example multilingual terminology extraction.

Sentence alignment is generally based on specific features of bi-texts: it is assumed that the translation generally follows the structure of the original text and that the sentences are usually chained in the same way in the source text and the target text. Furthermore, one can define a length ratio between two pairs of languages (for example, in terms of number of words, a French text is generally 1.2 times longer than the corresponding English text). The relative length of the sentences was the first criterion explored for sentence alignment. The first experiments in the domain of sentence alignment were made on the transcripts of the Canadian Parliament, since this corpus is of extremely good quality and the translation is very close to the original texts, contrary to what is usually found on the Internet.

Alignment Based on Relative Length of Sentences

A simple strategy for sentence alignment is to observe, first, that the sentences of a text vary in length, and second, that there is usually a good correlation between sentence length of the source text and sentence length of the target text. One can try to align the sentences on this basis; that is, by observing the relative length difference in the source text and looking for similar patterns in the target language. In order to avoid spreading alignment errors (i.e., a mistake in a given place that spreads to the rest of the text), it is therefore necessary to proceed somewhat globally, not just sentence by sentence. One way to solve the problem is to find specific patterns in the source language and observe whether the same patterns can be found in the target language. This way, it is possible to find "islands of confidence," or relatively reliable configurations distributed throughout the text.

Let's imagine a text composed of a given number of sentences and its translation. In the figure below, each cell is a sentence, and the number in each cell refers to the number of words in the sentence. Below is the source text:

8	13	12	7	14	10	8	13	5	22	12	11	14

And the target text:

9	16	13	23	12	9	20	23	14	6	7	15

Figure 4 Two texts of different length. Each cell with a number *n* corresponds to a sentence of length *n*.

We can see that the two texts do not have exactly the same number of sentences. The first three sentences have a relatively similar number of words and can therefore be linked together (note that the target language seems to systematically use a slightly larger number of words per sentence than the source language).

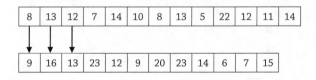

Figure 5 Beginning of alignment based on sentence length.

The same applies for the end of the text and some specific patterns in the text (for example, two consecutive sentences whose lengths are very different).

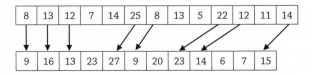

Figure 6 Other possible simple alignments.

Finally, the system tries to "bridge the gaps" by establishing links between the source and the target text, so as to obtain a fully connected bi-text in the end (each sentence in the source language must be linked to one or sometime two sentences in the target language). In the end, the system may have to proceed to "asymmetric alignments"; that is, connecting one sentence in the source text with more than one sentence in the target text.

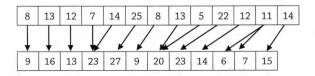

Figure 7 Alignment of remaining sentences.

Our example is clearly simplified. There are many ways of accomplishing a dynamic alignment: for example, by identifying the shortest and longest sentences or the difference in length between adjacent sentences; by calculating the length of groups of sentences in the first place; and so on.

Gale and Church (1993) applied this type of algorithm to the Hansard corpus (Canadian Parliament texts), obtaining an error rate of about 4% (i.e., 4% of the sentences were wrongly aligned). They show that this rate can even be lowered to less than 1% if only one-to-one mappings between sentences are taken into account (i.e., if we keep only single sentences in the source text that correspond to single sentences in the target text, which means that asymmetric alignments lead to more errors). They also show that the 1–1 relation corresponds to more than 89% of the sentences in the source text, about 9% correspond to 1–2 or 2–1 relations (i.e., one sentence in the source text is connected to exactly two sentences in the target language, or vice versa), and the other cases (i.e., a sentence that is not translated, or one sentence translated by three or more sentences) are very marginal.

The great advantage of this approach is its simplicity and its relative robustness. It has been proven that the method works well for different pairs of languages: the method is actually transferable and completely independent from the languages considered. It can even be applied to nonalphabetical languages with syllabic or ideographic writing, such as Asian languages. This robustness must nevertheless be qualified, since performance will worsen if the translation is not as reliable as in the case of the

Canadian Parliament transcripts, which is an exceptionally good corpus from this point of view. The method can also suffer from discrepancies (one misalignment leading to other misalignments, creating a cascading effect), even if the dynamic approach described above is meant to respond to this problem.

Different strategies have been devised to limit the problems of cascading misalignments. One way consists in trying to first find homogeneous text portions made of several sentences. Paragraphs are the most obvious units between the text and the sentence level, and paragraphs have been used with some success to complete the task. Additionally, most texts now come from the web, which means they contain HTML or other explicit tags that can be used for text alignment, since the target text may have the same structure as the source text. Finally, it is also possible to locate similar words in the original text and in the translation. This helps in finding what are called *correspondence points*. This approach is said to be *lexical*, since it is based on the analysis of a part of the lexicon.

Lexical Approach

Several studies have proposed strategies to align sentences based on lexical correspondences. This strategy is less generic than those described so far, but it is relatively efficient, especially between linguistically related languages.

If one considers a given bi-text (i.e., a pair of texts such that one is a translation of the other), one can often observe similar or nearly similar strings referring, for example, to person names, locations, and more generally proper nouns. These lexical correspondences are generally called *cognates*. Other elements can play a similar role, especially numbers, acronyms, and so on. Typography can also be helpful to identify related words, for example in bold and italics.

Xxx xxx xx xxxx xxx xxx.
Xxx xxx xx xxxxx xx x xxx xxx.
Xxx xxx xx xxxx xxxx xxx xx x xxx xxx.
Xxx xxx xx xxx xx x xxx xxx.
Xxx xxx xx Sarkozy xxx xx x xxx xxx.
Xxx xxx xx xxxx xxxx xxx xx x xxx xxx.
Xxx xxx xx xxxx xxxx xxxx xxx.
Xxx xxx xx xxxx xxxx xxx xx x xxx xxx.
Xxx xxx xx avril xxxx xxx xx xxx.
Xxx xxx xx xxxx xxxx xxx xx x xxx xxx.

Yy yyyy yyy yy y yyy yyy.
yyy yyy yyyy yyyy yyy yy y yyy yyy.
yy yy yyyy yyyy yyy yy y yyy yyy.
Yyy yy yyyy yyyy yyy yy y yyy yyy.
yyy y Sarkozy yyy yy y yyy yyy.
yyy yyy yy yyyy yyyy yyy yy y yyy yyyyy yyy yyyyyy yyyyy.
yyy yy April yyyy yyy yy y yyy yyy.
yyy yyy yy yyyy yyyy yyy yy y yyy yyy.

Figure 8   Two texts in a translation situation. Although the content of the texts is unknown (here represented by "xxx" and "yyy"), some words are identical or similar and can help determine reliable correspondence points.

All of these elements can be used to identify correspondence points between the source and the target texts. Sentence alignment is then calculated by resorting to dynamic programming, in a manner similar to what is done for alignment based on sentence length. Pairs of sentences with several correspondence points are most probably translations of each other. The process is applied iteratively until there is nothing left to align.

Mixed Approaches

It is, of course, possible to combine the two approaches in order to define a system based on both lexical indices and sentence length. On the one hand, cognates are rarely sufficient for aligning two texts. On the other hand, sentence length is generally a good feature for alignment, but it can happen that several consecutive sentences have a similar length. The idea is to find as many cues as possible between sentences to reinforce confidence in different local alignments.

Sentence alignment was a particularly active research topic in the 1990s. Researchers explored various cues, especially the structure of HTML documents, as seen above. The presence of titles, frames, and icons were used as features for the task. As a result of this research effort, the number of available bilingual corpora exploded in the 1990s. These new resources cleared the path for example-based translation and then for statistical translation, which is now the dominant paradigm in the field. The following chapters will discuss how these resources have been exploited in order to produce more robust and more reliable translation systems than those developed previously.

EXAMPLE-BASED
MACHINE TRANSLATION

Example-based translation, or translation by analogy, was introduced in the 1980s in Japan by Makoto Nagao (1984). Nagao noticed that traditional rule-based systems—still the common approach during the 1980s—tended to become more and more complex over time. As a result, they were progressively more difficult to maintain, which was of course a major problem. These systems generally also require a complete analysis of the sentence to be translated, which makes them extremely weak: if just one part of a given sentence cannot be analyzed, no translation will be provided for the whole sentence. Conversely, Nagao observed that professional human translators mainly work with fragments of text that they translate and recombine to form complete and coherent sentences. Translators generally do not carry out a complete preliminary analysis of the sentence to be translated, he argued.

At the same time, Nagao noticed that parallel corpora contain a great deal of valuable information that is for the most part lacking in bilingual dictionaries, even professional ones. Thus, he suggested, rather than trying to develop new dictionaries and new analyses or transfer rules between the languages at hand, it would be more convenient to directly use fragments of translation that one can find in existing bilingual corpora.

An Overview of Example-Based Machine Translation

Example-based machine translation typically operates in three stages to translate a given sentence:

• The system tries to find fragments of the sentence to be translated in the corpora available for the source language. All the relevant fragments are collected and stored.

• The system then looks for translational equivalences in the target language, thanks to the bi-texts used for translation.

• The system finally tries to combine the translation fragments to obtain a correct sentence in the target language.

Rather than trying to develop new dictionaries and new analysis or transfer rules between the languages at hand, it would be more convenient to directly use fragments of translation that one can find in existing bilingual corpora.

A simple example will illustrate this approach. Let's imagine that we ask the system to translate "*Training is not the solution to every problem*" into French and that a bilingual corpus is available with, among others, the following pairs of sentences (figure 9).

The system tries to find translational equivalents in the target language. For example, "*training is not the solution*" can be found in *Ex1* and *Ex2*. In both cases the translation

Ex1	*Training is not the solution to everything.*
	La formation n'est pas la solution universelle.
Ex2	*Training is not the solution to all parenting struggles*
	La formation n'est pas la solution à toutes les difficultés rencontrées par les parents.
Ex3	*There is a solution to every problem.*
	Il y a une solution à tous les problèmes.
Ex4	*There is a spiritual solution to every problem.*
	Il y a une solution spirituelle à tous les problèmes

Figure 9 Automatically extracted sentences from a bilingual corpus in order to translate the sentence "*training is not the solution to every problem.*" Each sentence in English contains a sequence of *n* similar words with the sentence to be translated.

includes "*la formation n'est pas la solution*": the system can infer that it is a translation of the English sequence, since this expression is shared by the two sentences in the target language. Likewise, from *Ex3* and *Ex4*, the system can infer that "*to every problem*" can be translated into French as "*à tous les problèmes*." By combining these two identified sequences of words, the system produces the translation "*la formation n'est pas la solution à tous les problèmes*."

As one might imagine, this is a very simplified example, one where the sentence in the source language directly corresponds to long existing sequences of words in the target language. In practice, the problem is obviously more complicated.

The Search for Translation Examples

Since it is very rare to find exact matches at sentence level between the text to be translated and the available bilingual corpus, it is necessary to find equivalences (called examples in this approach) at the infra-sentential level, as seen in the previous example. But even at the infra-sentential level, searching for translational equivalences is a complex problem: (i) the identified "examples" (sequences of exactly matching words) are often very short; (ii) for the same sequence in the source languages, different translations can often be found, and it is not obvious how to choose the most relevant one; and (iii) merging different fragments of text is difficult, since fragments often overlap or are not fully compatible with each other. Thus, rather than just looking for exact equivalences at word level (or character level), it is useful to try to find equivalences on a more general basis in order to make the approach more robust. This is often referred to as *translation by analogy*: it no longer involves finding fragments of texts that reproduce the exact sentence to be translated, but rather fragments of texts bearing an analogy with the sentence to translate.

Several techniques have been proposed to find analogies, translational equivalents, or "examples" on a more or less linguistic basis:

- comparison of strings of characters,

- comparison of words,

- comparison of sequences of linguistic tags (i.e., noun, verb, etc.),

- comparison of linguistic structures.

The first approach, comparison of strings of characters, has the advantage of being independent of the languages considered and can, for example, also be applied to Asian languages (the fact that a word can often be a single character in Asian languages is not a problem for this approach). The second strategy, comparison of words, is well suited for languages with good lemmatizers (i.e., tools able to recognize words as they appear in a dictionary), but this is not the case for all languages. Moreover, as already said, these approaches are too close to the surface of the text. There are too many sources of variation in a language to make these techniques really powerful.

More advanced techniques rely on a stage of text preprocessing to enrich available bilingual corpora with higher-level information. In practice, linguistic tags are added to words in order to make it possible for the system to have a more abstract representation of the data. The preprocessing stage generally includes part-of-speech tagging (recognizing adjectives, nouns, verbs, etc.), and sometimes a shallow semantic analysis (recognizing dates, proper nouns, idioms, etc.). Transfer rules mapping linguistic sequences between the source language and the target language must then take into account this new information. For example, if adjectives have been described as optional, *"there is a spiritual solution"* can be used to translate *"il y a une solution,"* even if the French fragment does not include any word related to *"spiritual."* Of course, when linguistic equivalences are not perfect, the translation may possibly be quite different from the original text.

Lastly, the recognition of specific syntactic structures would make it possible to proceed through a direct comparison of *syntactic trees* (i.e., a representation of the structure of the sentence). In theory, this strategy makes it possible to compare sentences that look very different at surface level (i.e., if one just looks at sequences of words). For example, the two sentences *"he gave Mary a book"* and *"he gave this book to Mary"* have the same syntactic structure, although a system that looks only at sequences of words would find merely similar fragments ("he gave" and a few isolated words).

Once the relevant fragments in the target language have been collected, a series of rules or statistical indices are applied to try to recompose a complete sentence from the identified fragments. This is a difficult task because these fragments are usually partial and incomplete, and they overlap and do not correspond to autonomous syntactic phrases. Some research groups tried to develop systems using only relevant syntactic phrases (complete noun phrases or verb phrases, for example) but this does not result in any improvement for different reasons, mainly data sparsity (it is very difficult to collect enough relevant examples at phrase level).

Appeal and Limitations of Example-Based Machine Translation

Example-based machine translation generated great interest during the 1980s. Rather than developing a machine translation system manually, which is long and very costly, the example-based approach allowed for optimal exploitation of large quantities of bilingual texts that were beginning to be available at the time. It is clearly not a coincidence if this approach emerged at the same time as the first work on bilingual text alignment.

The approach was mainly explored for Asian languages that do not show the same similarity as, for example, French-English pairs. Hence, the Japanese structure *Noun1 no Noun2* (*Noun1 の Noun2*) is often cited as an example, because the particle "*no*" (corresponding to the Japanese character の) can represent various types of links between two nouns. Here are some examples frequently cited in the literature and taken from an article from 1991 (Sumita and Iida, 1991):

Mitsu **no** hoteru	Three hotels
Isshukan **no** kyuka	A week**'s** holiday
Kyouto **deno** kaigi	The conference **in** Kyoto
Kaigi **no** sankaryou	The application fee **for the** conference
Kaigi **no** mokuteki	The objective **of** the conference
Youka **no** gogo	The afternoon **of** the 8th

Figure 10 Different examples with the Japanese particle "no." One can see that the particle requires the use of a different linguistic structure each time when translating into English, depending on the context (see Sumita and Iida, 1991).

It is clear from figure 10 that "*no*" can express a wide variety of possible relations between the two nouns: it can be a kind of genitive, but it can also express indications of goal, time, or location. The authors demonstrate that it is hardly possible to formalize this by rules, since it would require the system to have access to semantic information. The example-based approach seems to be more appropriate, as long as the examples provide a good coverage of the text to be translated.

The limits of this approach are clear: by default, if no translational fragment is found from the set of examples, the system will either fail or produce a word-for-word translation. This approach was essentially explored for genetically distant languages (typically Japanese-English) for which it seemed difficult to develop manual transfer rules. Rather than describing specific contexts manually, the proponents of the example-based approach observed that an ambiguous pattern can be disambiguated with a proper look at semantic classes and explicit markers in the translational equivalences.[1] For example, in the case of the Japanese particle "*no*," it is possible to find equivalences with the English genitive marked by "'*s*" and with sequences introduced by specific prepositions ("*for*," "*in*," "*of*," etc.). Each of these elements ("*no*" on one hand, and genitive markers or prepositions on the other hand) is considered as a "marker."

The approach was also used for specific domains using a particular sub-language with a limited vocabulary and a very specific and highly regular terminology and phraseology. This is, for example, the case of computer documentation, which is one of the main fields where example-based translation has been tested with some success (Somers, 1999; Gough and Way, 2004). In such contexts, sentences are regular and the same expressions are often used, which means that the example-based approach can obtain an acceptable coverage of the text to be translated. The main problem remained, however, extending the coverage, which is always partial, even for the most regular texts. The consequence is that example-based translation is interesting but can hardly be used alone in practical contexts.

Example-based translation has therefore sometimes been used as a module within a more complex system. Mixing the example-based approach with a statistical analysis of very large corpora has proven to lead to very interesting results, since statistical approaches are known to have good recall and can in turn benefit from the precision of the example-based paradigm.

STATISTICAL MACHINE TRANSLATION AND WORD ALIGNMENT

Since the late 1990s, aligned bilingual corpora have been the object of various studies aiming to extract translational equivalencies between languages at the word or phrase levels. For example, a popular task of that decade was to extract bilingual lexicons for human translators from available parallel texts. Attempts to produce entirely automatic translation systems through statistical analysis took place simultaneously—the approach that remains most popular today and has given rise to the most pioneering research in the field.

Word alignment is a considerably more complex task than sentence alignment, as one can imagine. While there is often a 1–1 correspondence between the source text and its translation at sentence level (in other words, one sentence from the source text most often corresponds to one sentence in the translation), this does not necessarily apply at word level. It is well known that languages differ significantly and that many words cannot be directly translated. Most correspondences are therefore said to be "asymmetrical"; that is, one word from the source or target language corresponds to 0, 1, or n words in the other language.

Some Examples

Consider the following example: "*Thanks to those in the field for their insights*," translated as "*Merci à tous ceux qui, sur le terrain, ont fait part de leurs idées*" (taken from the website www.unaids.org). The English sentence contains nine words, whereas the French equivalent contains 14! It is thus difficult to propose an alignment at word level, since the two sentences do not have the same structure (the French version introduces a relative clause, whereas English uses simpler and more direct wording). Figure 11 presents an example of an incomplete lexical alignment

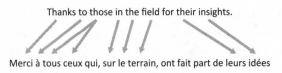

Thanks to those in the field for their insights.

Merci à tous ceux qui, sur le terrain, ont fait part de leurs idées

Figure 11   A possible alignment between two sentences.

between the two sentences (incomplete in the sense that some words do not have translational equivalents).

In this example, the equivalent of the preposition "*for*" is in some ways "*qui ont fait part de*": we could imagine relating "*for*" to all the words in the French expression, but this would hardly make any sense. This is also very true for the following example.

In figure 12, several words from the source sentence are not directly translated in the target sentence. Furthermore, we see that links between the two sentences intersect several times due to the inversion of the two propositions (the English sequence "*...that what he has announced he will actually do...*" appears in French as "*... qu'il fasse réellement ce qu'il a annoncé*"). One word from the source language can furthermore correspond to several words in the target language, and vice versa ("*will see*" corresponds to the simple verb form in the future tense "*veillerons*," whereas, in the opposite direction, a link can be established between the simple word "*what*"

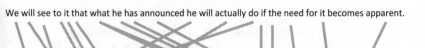

Figure 12 A possible alignment between two sentences, with several intersecting links.

and the French expression "*ce que*"). Lastly, the end of the sentence is difficult to "align" correctly: it would be more satisfactory to draw a link directly between the phrase "*the need for it becomes apparent*" and "*la nécessité s'en fait ressentir*," because it is hard to see how this very good translation could be decomposed (the phrase "*the need*" can probably be aligned with "*la nécessité*," but nothing in French can be considered a direct translation of "*becomes apparent*").

In sum, identifying lexical equivalences is a difficult task to automate because the "search space" (i.e., the number of possibilities to be considered) is huge: each word from the source language can potentially be linked to any single word or group of words in the target language. This is, of course, not true when a human being is performing the task on known languages, but imagine how complex it would be with completely unknown languages! This is the case for computers, which do not have any idea of syntax or semantics and do not have access to a lexical resource like a dictionary. From a linguistic point of view, the task seems even more questionable, since one thing we know about translation is that there are no direct equivalences between languages at word level. The proof is that a word-for-word translation is generally very bad. For the most part, these ideas are right, and we will see that for the past several years, efforts have focused on taking into account more complex sequences of words in machine translation in order to avoid the basic errors that arise from the word-for-word approach.

Nonetheless, toward the end of the 1980s, the statistical approach based on sentence alignment at the word level led to remarkable progress for machine translation. This approach naturally takes into account the statistical nature of language, which means that the approach focuses on the most frequent patterns in a language and, despite its limitations, is able to produce acceptable translations for a significant number of simple sentences. In certain cases, statistical models can also identify idioms thanks to asymmetric alignments (one word from the source language aligned with several words from the target language, for example), which means they can also overcome the word-for-word limitation.

In the following section, we will examine several lexical alignment models developed toward the end of the 1980s and the beginning of the 1990s. The goal of this approach is to use very large bilingual corpora to automatically extract bilingual lexicons. In these lexicons, different translations are proposed for each word, and each of these translations is assigned a score reflecting its probability of being a correct translation. These lexicons are a fundamental part of machine translation systems, as they provide the basis for a word-for-word translation.

The "Fundamental Equation" of Machine Translation

At the end of the 1980s, an IBM research team located in Yorktown Heights, New York, decided to develop a machine translation system based on techniques initially developed for speech transcription. Speech transcription refers to the task of producing a written text from a sound sequence. Translation can be seen as a similar task, the only difference being that the input signal is a sequence of words in the source language instead of a sound sequence.

The IBM experiments were described in a series of papers published at the end of the 1980s and the beginning of the 1990s (see Brown et al., 1988, 1990 and 1993). The authors take as a starting point the fact that there are always several possible translations for a given sentence, whatever the source and target language may be. The choice among these various possibilities is to some extent a matter of taste and personal choice. Bearing this in mind, one can consider that any sequence in the target language can be considered a translation, to a certain extent, of a sequence in the source language. Given a pair of sentences (S,T), where S is the source sentence and T the target sentence, it is possible to calculate a probability $\Pr(T|S)$ that a human translator would produce translation T from the sequence S. The idea is that $\Pr(T|S)$ will be very small for a pair of sentences like (*Le matin je me brosse les dents* | *President Wilson was a good lawyer*) and a lot higher for a pair such as (*Le président Wilson était un bon avocat* | *President Wilson was a good lawyer*). In other words, every translation in the target language can be considered a translation of a sentence in the source language, but realistic translations will obtain a score well above 0, whereas the others will remain close to 0.

The IBM team showed that this hypothesis can be modeled using well-known principles from probability theory, namely Bayes' theorem. Bayes' theorem in a way reverses the problem and aims to determine, given various sequences from the target language, which has the most chance of being a translation of the source language. This can be formalized with the following equation:

$$\Pr(T|S) = \frac{\Pr(T)\Pr(S|T)}{\Pr(S)}, \tag{1}$$

where $\Pr(T)$ is a language model of the target language and $\Pr(S|T)$ a translation model. In other words, $\Pr(S|T)$ measures the probability of sequence S according to sequence T (meaning the probability that S is a sequence in the source language that corresponds to sequence T; if the probability is close to 1, then the two sentences are probably a translation of one another), whereas $\Pr(T)$ measures the probability of the sequence in the target language, without taking the source language into account (i.e., the probability that T forms a valid and well-formed sequence in the target language, thus accounting for word order in the target language).

For example, Pr(T) encodes the fact that the sequence "*the red car*" is more frequent than "*car the red*," "*car red the*," or "*red car the*." The translation of "*la voiture rouge*" will then be "*the red car*" rather than any other sequence formed with these same words. We will see later that the translation model Pr(S | T) makes the translation process possible by breaking down the sentence into small fragments to search for equivalences at word level. When all word-for-word equivalences are put together, different sentences are possible that differ only in their word order. Pr(T) helps "select" among these solutions the sequence that has a chance of being the most correct in the target language, taking into account only considerations of word order.

Since the denominator of equation (1) does not depend on T, the equation can be simplified as follows:

$$T' = argmax_T \, [\Pr(T) * \Pr(S \, | \, T)] \tag{2}$$

For the IBM team (see Brown et al., 1993), this formula is the "fundamental equation of machine translation" because all statistical models thereafter originate from it.

It should be noted that the equation does not explain how to decompose the source sentence during the translation process. The easiest way is for the model to rely on words and perform a word-for-word translation, at least in a first approximation. In this context, the idea is to find, for each word in the source language, an equivalent in the target language using a very large aligned corpus at sentence level, as we saw in chapter 7. In order to do this, the IBM team proposed to decompose the statistical translation process into three different steps:

1. determine the length of the target sentence depending on the length of the source sentence;

2. identify the best possible alignment between the source sentence and the target sentence; and then

3. find correspondences at word level (i.e., find word m_t in the target language corresponding to word m_s in the source language).

This strategy clearly gives a very simplified picture of the translation process. In particular, the first step assumes that every sentence of length l in the source language will be translated by a sentence of length m in the target language! In fact, at the end of the 1980s, IBM was fully aware of the limitations of the approach: various articles published at the time stressed the fact that such an approach would have to be complemented by efforts to incorporate more linguistic knowledge and more complex matching rules between languages. For the IBM team it was even probable that this approach would not make it possible to go very far given its inherent limitations. But IBM sought to evaluate the quality of the results obtained with a very simple approach compared to more complex systems obtained after years of human effort. We will soon see that the results obtained were incredibly good from this point of view.

The essential part in the IBM model thus resided in the choice of words for the translation in the target language, which means that the essence of the overall model can be found within the lexical alignment strategy (i.e., alignment at word level). The overall approach comprises two very different steps. The first one aims at extracting as much information as possible from very large bilingual corpora; the second one uses this knowledge to translate new sentences. More precisely, the approach can be described as follows:

1.   A word alignment algorithm is applied to a very large corpus made of bi-texts (texts aligned at sentence level). The result of this analysis is twofold: a bilingual dictionary (i.e., the result of the alignment at word level) and the most likely global alignment at sentence level.

2.   This huge amount of information is then used to translate new sentences that the end user wants to translate.

The first step is often called the "training step" or "learning step," and the second step the "processing step" or "testing step." It is important that the data used for training be similar to the data used for testing in order for the system to produce satisfactory results. As one can imagine, the key point lies in the quality of the information accumulated during the training step, which essentially entails analyzing a very large aligned corpus at word and sentence levels. The seminal paper from IBM in 1993 described five alignment models, each of which is a modification of the previous model.

Different Approaches for Lexical Alignment: The IBM Models

As we have seen, the translation approach developed within IBM in the late 1980s was essentially based on translation choices carried out at word level. So, the crucial ingredient in this approach is an accurate bilingual dictionary. With the statistical framework, a bilingual dictionary consists in fact, for each word, of a list of possible translations in the target language, along with a probability associated with each of these possible translations. For practical reasons, the sum of the probabilities of all the possible translations associated with a given word in the source language must equal 1, as shown in table 1.

Table 1 Example of possible translations in French for the English word "*motion*"

English word	Possible translation	Probability
motion	mouvement	0.35
	geste	0.12
	motion	0.11
	proposition	0.10
	résolution	0.10
	marche	0.05
	signe	0.04
	...	...
		Total = 1

Note: Each translation appears with a probability based on the number of times the word was actually translated in this way (compared to the total number of occurrences of the word in the corpus). Thus, "*mouvement*" is the most likely translation, followed by "*geste*," etc. This list has been limited arbitrarily to the first seven possible translations, but it is theoretically possible to list as many as necessary, as long as the sum of the probabilities ultimately equals 1 for one given word in the source language.

The different IBM models are in fact not limited to 1 to 1 correspondences at word level, but assume that a source word may be aligned with 0, 1, or *n* target words. The different models (numbered 1 to 5) include different optimizations to deal with multiword expressions in the target language, or with words with no equivalent in the other language (for example, determiners that appear in one language but not in the other one). We give a quick overview of the various models below, without all the mathematical details.

Model 1

The first model developed by IBM was extremely simple. It considered that initially, by default, any word from the target language could be the translation of any word in the source language (within two sentences that are translations of each other, taken from a given bi-text). This starting point may seem too crude, but one should keep in mind that the system initially does not have any linguistic knowledge (no dictionary is provided) and will base its analysis on very large corpora (several millions of aligned sentences are used in most systems nowadays). To illustrate the approach, in what follows, we take the example of a single, isolated sentence, but it should be borne in mind that the approach can only work if regularities are identified from millions of examples.

In order to roughly determine the probability that a target word m_t is the likely translation of a source word m_s, one could collect all the words appearing in all the translations of sentences where m_s appears and then calculate the translation probability of each word from the relative frequencies of the words collected this way. This means, intuitively, that in the absence of any linguistic knowledge, the system assumes that all the words in the target sentence are possible translations of all the words in the source sentence. Thus, for the sentence pair *"the cat is on the mat"* ⇔ *"le chat est sur le paillasson,"* the six French words "le," "chat," "est," "sur," "le," and "paillasson" will be considered as equally probable translations of "cat," as of any other word in the English sentence. Clearly, this strategy does not work for one sentence in isolation (*"paillasson"* is not a proper translation for the English word "cat"), but the analysis of a huge number of sentences will reinforce the association *"chat"* ⇔ *"cat"* (since these two words very frequently appear together in bi-texts), whereas the association between "cat" and "paillasson" will remain very marginal (meaning that in the end the association will have a probability close to 0) and as a result will likely be ignored thereafter.

However, this simple solution poses a major problem: if the target sentence contains 20 words, each of them will be a possible translation that will have the same weight as if the target sentence contained five words. Yet, as there is only one single equivalent m_t to find for each m_s in this case, it is obvious that a longer sentence will lead to more noise (meaning it will generate more erroneous possibilities) than a shorter sentence. In other words, the number of words in the sentence should be taken into account in order to increase the probability of each of the five words in the shorter sentence.

Simultaneously, the system also considers the global probability of all word alignments at sentence level: the reinforcement of some connections at word level (like between "cat" and "chat") will reinforce some possibilities at sentence level, and vice versa, as will be described below.

Following this principle, IBM defined a process that uses a classic learning algorithm called the *expectation-maximization algorithm*, or EM, which gradually calculates the probabilities associated to each pair m_s-m_t, as well as the probabilities associated to each possible alignment at sentence level. As already noted, these two probabilities depend on and gradually reinforce each other. The EM algorithm calculates this joint probability in two steps: (i) arbitrary initial values are first assigned to each of the parameters (typically, every word in the source sentence can be connected with every word in the target sentence with the same probability); (ii) the system then calculates in an iterative manner the probabilities of the overall alignment at sentence level and then again at word level until convergence is achieved (the process is iterative since the alignment at sentence level changes the probabilities at word level, and vice versa, until the system reaches a stable state).

Let us take an example. Each alignment and each lexical correspondence has the same probability at the beginning. The fact that two words appear regularly together in bi-texts (in a source sentence and in the corresponding target sentence) will gradually strengthen their probability of being a translation of each other, as well as the probability of possible alignments at sentence level where these two words are connected. The figures below (figure 13 to figure 16, based on Koehn, 2009) demonstrate clearly the alignment process at word level.

The reader is referred to IBM's initial publication (Brown et al., 1993) for all the mathematical details. The application of such an algorithm to very large corpora requires being able to control memory management and complex computational methods. Moreover,

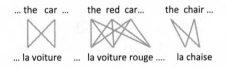

Figure 13 Initialization of the alignments. Each English word is linked with equal probability to all the words in the French translation.

Figure 14 After the first iteration, the algorithm identifies the link between "la" and "the" as being the most likely, based on their frequency in the source language and in the target language. These links are strengthened (shown in bold) to the detriment of other links and therefore also other possible alignments.

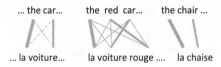

... the car...   the red car...   the chair ...

... la voiture...   la voiture rouge ....   la chaise

Figure 15   After another iteration, the algorithm identifies the other most probable links between "voiture" and "car," then between "chaise" and "chair" and between "red" and "rouge." The other possible links and alignments become less and less probable.

... the car...   the red car...   the chair ...

... la voiture   ...   la voiture rouge ....   la chaise

Figure 16   The process ends when there is convergence, meaning a stable structure has been found. The other links in the figure are removed, but in fact they remain available with a very low probability. It is possible to filter the alignment using a threshold in order to select only a limited number of possibilities, as shown in this figure, where the alternative links have been completely deleted.

our presentation is simplified here: in practice, not only 1–1 word correspondences are considered but also 1 to m (when one word in the source language corresponds to m words in the target language) which makes the problem even more complex. Finally, it should be noted that the algorithm guarantees that the result is optimal (i.e., there is convergence and the system always stops after a certain number of iterations, which is not always the case by default with the EM algorithm).

The other models are all derived from this initial model. They make it more complex in order to take better account of certain language particularities and to provide better translations.

Model 2

As we have seen, IBM Model 1 considers that all initial alignments at word level have an equal probability (i.e., all the source words m_s can be linked to all target words m_t with an equal probability at the beginning of the alignment process). This is clearly wrong: one can easily observe that word order is often roughly similar between different languages, especially between typologically related languages (such as French and English). This is, of course, not always true, but there is nonetheless a strong correlation between the word order in the source and target languages in this case.

Model 2 thus modifies Model 1 by taking into account in its calculation the relative position of word m_t in relation to the source word m_s. This does not fundamentally change the previously mentioned algorithm, but leads to better results and speeds up the learning process before reaching convergence.

Model 3

Model 3 is notably more complicated than Model 2. Its goal is mainly to better formalize the question of the 1-n

correspondences (where one word in the source language is translated with many words in the target language; see, for example, *"potato"* in English, which corresponds to *"pomme de terre"* in French). This issue is not addressed by the previous models but is very common when translating between any languages. IBM Model 3 also addresses other related problems: the article *"the"* is often translated as one word in French (*"le," "la," "les," "l'"*) but is also often omitted; *"only"* may be translated as *"seulement"* but also by the expression *"ne... que"* (two non-contiguous words), etc.

The IBM team thus proposed to add to Model 2 "fertility probabilities" that indicate, for each word in the source sentence, the possible number of words in the target sentence (by default, each single word is translated by another simple word, but for *"potato,"* the translation is *"pomme de terre,"* which means there are systematically three words in French in this case; for *"the,"* 0 or 1 word can be generated, etc.).

A related process aims at extrapolating a number of semantically empty words in the target sentence. This solves one of the limitations of previous models, since they always need a word in the source language in order to be able to generate a word in the target language. For example, to translate *"il est avocat"* into *"he is a lawyer,"* the English article *"a"* must be added, which was not possible with the previous model (IBM Model 2). This problem is solved through "distortion probabilities," which allow for empty positions during alignment in order to properly generate these new words in the target language.

Model 4

The IBM team then observed that some parts of a sentence can be moved more or less freely, and that this may be the cause of structure variation between a sentence and its translation. A pair of sentences may have the same structure in the source and the target language, but differ because one phrase has been moved (see, for example, *"He has lived in New York since last year"* vs *"Il habite depuis l'année dernière à New York"*: the two sentences have the same structure, except that *"in New York"* / *"à New York"* is at the end of the sentence in the French version).

The previous models, mainly based on correspondences at word level, were not very good at tackling this kind of problem. Model 4 therefore modified the distortion probabilities proposed in Model 3 to account for these blocks of text that can move within the sentence.

Model 5

Model 5 did not make any fundamental modifications to Model 4, but made it possible to avoid irrelevant sequences that would be otherwise considered given the way the problem is formalized. Model 5 is more accurate mathematically, but the calculations required are in fact much more complicated and the results are similar, or even worse than those from the previous model since Model 5 requires more training data. In brief, this model can be ignored here, as it mainly concerned calculation issues and added nothing new from a linguistic point of view.

The Translation (or Processing) Stage

At this point, we need to remind the reader that with the statistical approach, the translation process consists of two fundamental steps. The system first uses a very large corpus of bi-texts (texts aligned at sentence level) to automatically acquire information about the translations of words and about the possible alignment of all the words at sentence level. This is what we have described in the previous section (this phase is often called the "training" or "learning" phase, or even "encoding phase," since this stage involves encoding information about the language).

The knowledge extracted from the bilingual corpus is then used during the processing phase (also known as the "test" or "online" phase) to translate new sentences submitted to the system. This stage is also known as the "decoder," since the system tries to "decode" the input sentence as one would decode a secret message.

Each time a new sentence in the source language is submitted as input, the system splits the sentence into words, looks for the most likely translation for each word, and takes into account the constraints on word order as given by the translation model.

The language model makes it possible to evaluate the probability of different candidate translations in the target language. This is crucial for the quality of the result: thanks to the language model, it is possible to consider translations that are not necessarily based on the most probable equivalences at word level. The translation of a sentence made only by concatenating the most probable translation of each single word may have a very low probability as a whole, whereas a sequence of words including translational words with a lower probability may have, at sentence level, a higher probability. Let us take an example. The sentence *"The motion fails"* contains the word *"motion,"* for which the most probable translation is *"mouvement"* (see the previous section). By default, a translation (based uniquely on the most probable words) would therefore be *"le mouvement est rejeté,"* which does not mean anything in French. The translation *"la motion est rejetée"* is much more probable (even if *"motion"* is less probable than *"mouvement"* for the English word "motion"). This translation is the correct one, which is rightly predicted by the model.

Finding the best translation actually involves sorting through a multitude of possible choices, each word having multiple translation equivalents, or even not being

Finding the best translation actually involves sorting through a multitude of possible choices, each word having multiple translation equivalents, or even not being translated into the target language.

translated into the target language. The module known as the "decoder" is responsible for finding the best translation possible. Its role is to find the solution that maximizes the score at sentence level, taking the translation model and the language model into account.

The techniques employed to solve this kind of problem can be relatively complex from a computational point of view. The goal is to gradually eliminate the least probable local hypotheses to efficiently converge toward the most probable global solution. This type of algorithm is not specific to machine translation, but is already frequently used for speech analysis. As for speech, finding a good translation involves calculating an optimal score from a very large number of partial analyses that overlap and are often incompatible with one another.

Back to the Roots of the Domain?

The IBM models are in some ways a return to the roots of the domain, since the techniques proposed very directly echo several of Weaver's 1949 propositions. The fact that the module aiming at producing the translation is sometimes called a "decoder" is no coincidence. This is a reminder of the general model of communication, and the goal is to "decode" the source text by translating it into the target language (see chapter 5).

The models designed by IBM achieved phenomenal success. They were revised, modified, and improved. They remain the basis of most machine translation systems used today, although all the major players in the field are now moving towards a new kind of techniques called deep learning (see chapter 12).

However, these models also have their own limitations, the main one being that they require huge quantities of data to achieve reasonable performance. In the following chapter, we will take a look at recent developments around these models, but also at the situation of rare languages for which not enough data are available to achieve accurate statistical translation systems.

SEGMENT-BASED MACHINE TRANSLATION

We saw in the previous chapter the success of the models developed by IBM for machine translation. One of the main limitations of these models is the fact that they are mainly based on alignment at word level (i.e., they mainly produce word-for-word translations, even if they also allow 1-m alignments, where one word in the source language corresponds to several words in the target language). This chapter covers developments that took place in the 1990s and 2000s that aimed to overcome the main limitations of the IBM models. We examine how information of a syntactic and semantic nature has been progressively integrated into models to compensate for the limitation of purely statistical approaches.

Toward Segment-Based Machine Translation

The IBM models have been subjected to numerous enhancements. The most significant improvement was to take into consideration the notion of segments (or sequences of words) in order to overcome the limitation of a simple word-for-word translation. Among the other improvements, the notion of double alignment is worth mentioning, since it greatly increases the quality of the search for translations at word level in bilingual corpora.

Double Alignment

The IBM models are able to recognize correspondences such that, for one word in the source language, there is 0, 1, or n words in the target language. However, the original IBM models, because of their formal basis, do not make it possible to obtain the opposite correspondences (in other words, one word from the target language cannot correspond to a multiword expression in the source language, for example). This is a strong limitation of these models that has no linguistic basis, since multiword expressions clearly exist in every language. It therefore seemed necessary to overcome this limitation imposed by the IBM models in order to allow for m-n alignments (where any number of words from the source language corresponds to any number of words in the target language).

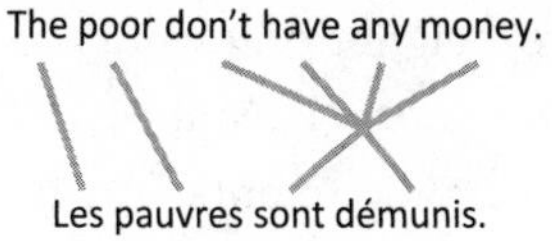

The poor don't have any money.

Les pauvres sont démunis.

Figure 17 Example of an alignment that is impossible to obtain from IBM models. The sequence *"don't have any money"* corresponds to the group *"sont démunis"* in French: this is an example of an *m-n* correspondence (here, *m*=4 and *n*=2 such that four English words correspond to two French words, if we consider "don't" as a single word).

The original IBM article specifically mentioned the example shown in figure 17, which the models proposed in 1993 were unable to handle.

One way of overcoming this problem is to first calculate the alignments from the source language into the target language, then repeat the operation in the opposite direction (from the target language into the source language). The shared alignments are kept, namely those concerning words that have been aligned in both directions. The alignments obtained using this technique are generally precise but provide a low coverage of the bi-texts. Globally, this method has two major defects: first, the process is more complex than a simple alignment, and is therefore more costly in terms of computation time; second, at the end of the process, a large number of words are no longer aligned because the constraints imposed by the double-direction analysis cause numerous alignments identified in a single direction to be rejected. Various heuristics then have to be used to expand the alignments to neighboring words in order to compensate for the coverage problems incurred (the double alignments, or "symmetric alignments," can be seen as "islands of confidence"; see chapter 7).

It has been shown that this method improves the results of the original IBM models. However, in order to obtain a good coverage of the data with these models, it is necessary to have huge quantities of data, which makes them impractical in some circumstances.

The Generalizations of Segment-Based Machine Translation

We have seen that the double alignment approach helps to identify *m-n* translational equivalences, such as *"don't have any money"* ⇔ *"sont démunis"* (where four English words amount to two French words). In fact, it is possible to generalize the approach so to consider the problem of translation as an alignment problem at the level of sequences of words, and not at the level of isolated words only. The goal is to translate at the phrase level (i.e., sequences of several words): this would enable the context to be better taken into account and would thus offer translations of better quality than simple word-for-word equivalences.

Several research groups have tackled this problem since the late 1990s, and various strategies have been explored. One strategy is to systematically symmetrize the alignments (see the previous section) in order to identify

It is possible to generalize the approach so as to consider the problem of translation as an alignment problem at the level of sequences of words, and not at the level of isolated words only.

all the possible *m-n* alignments. Other researchers have tried to directly identify linguistically coherent sequences in texts, through rules describing syntactic phrases for example (this can be seen as a first attempt to introduce a light syntactic analysis in the translation process). A last line of research tried to import some techniques from the example-based paradigm (see chapter 8), the idea being to make the alignment process both more robust and more precise by aligning from tags and not from word forms. For example, the following two sentences may seem very different for a computer, since several words are different: "*In September 2008, the crisis ...*" and "*In October 2009, the crisis ... .*" However, if the system is able to recognize date expressions, it is possible to recognize the structure "In <DATE>, the crisis" in both sentences: they can thus be aligned successfully. This technique can significantly improve the quality of the alignment.

The results obtained by these models show a clear improvement in comparison to the more complex IBM models, notably IBM model 4. However, the results are still very dependent on the training data: the more data there are, the more accurate the models will be. Moreover, segment models require a lot more training data than models based only on word alignment. Finally, it should be noted that the notion of segments does not generally correspond to the notion of phrases. A closer look at the results obtained shows that the segments obtained by training from large bilingual corpora correspond to frequent but fragmentary groups of words (for example, "*table of*" or "*table based on*"). On the contrary, limiting the analysis to linguistically coherent phrases (for example, "*the table*" or "*on the table*") seriously affects the results. In other words, if one forces the system to focus on linguistically coherent sequences corresponding to syntactically complete groups, the results are not as good as with a purely mechanical approach that does not take syntax into account.

The most challenging part of segment-based translation is for the system to produce a relevant sentence from scattered pieces of translation. Figure 18 gives a simplified but typical view of the situation after the selection of translation fragments (this view is simplified, because here the sentence to be translated is short, the number of segments to be taken into account is limited, and in real systems all fragments have a probability score).

It is clear from figure 18 that only the careful selection of some fragments can lead to a meaningful translation. The language model of the target language helps in finding the most probable sequence in the target language; in other words, it tries to separate linguistically correct sentences from incorrect ones (independently from the source sentence at this stage).

As one can easily imagine, these models are much more complex than the original IBM models based on

Figure 18 Segment-based translation: different segments have been found corresponding to isolated words or to longer sequences of words. The system then has to find the most probable translation from these different pieces of translation. It is probable that *"les pauvres n'ont pas d'argent"* will be preferred to *"les pauvres sont démunis,"* but this would be acceptable since the goal of automatic systems is to provide a literal translation, not a literary one.

simple words. Thus, they may require considerable processing time compared to the original IBM models. The increasing computational power of computers somewhat compensates for this problem. From a linguistic point of view, it should be noted that these models fail to identify discontinuous phrases (where one word in the source language corresponds to two noncontiguous words in the target language), which are crucial in languages such as French or German (English: "*I bought the car*" ⟺ German: "*Ich habe das Auto gekauft*"; English: "*I don't want*" ⟺ French: "*Je ne veux pas*").

The recent developments we have described in this section have, however, helped improve the IBM models and can still be considered currently as the state of the art in machine translation.

Introduction of Linguistic Information into Statistical Models

Statistical translation models, despite their increasing complexity to better fit language specificities, have not solved all the difficulties encountered. In fact, bilingual corpora, even large ones, remain insufficient at times to properly cover rare or complex linguistic phenomena. One solution is to then integrate more information of a linguistic nature in the machine translation system to better represent the relations between words (syntax) and their meanings (semantics).

Alignment Models Accounting for Syntax

The statistical models described so far are all direct translation systems: they search for equivalences between the source language and the target language at word level, or, at best, they take into consideration sequences of words that are not necessarily linguistically coherent. As we have seen, a strategy that takes into account the fragments identified on a purely statistical basis is more efficient than a strategy that only retains linguistically motivated phrases (e.g., noun phrases or verb phrases).

Several attempts have been made, nonetheless, to take better account of syntax in the machine translation process. This can be done through the integration of parsers, also known as syntactic analyzers: a parser is a tool that tries to automatically identify the syntactic structure of the sentences to be analyzed. If we recall Vauquois' triangle (see figure 2 in chapter 3), syntax makes it possible to take into account the relations between words. For example, there may be a relation between distant words in the sentence that is very difficult to spot with a purely statistical model. Parsers, however, are theoretically able to consider the relations between the words, even in cases where those words are not adjacent on the surface. In Vauquois' words, syntax involves transfer rules between equivalent syntactic structures, not a word-for-word translation.

From a theoretical perspective, this type of approach can better analyze discontinuous morphemes that may be common in some languages (as in the example in German above, "*Ich habe das Auto gekauft*," where the analyzer should identify "*habe gekauft*" as a whole despite the distance between the two words). Syntax can also help represent the connection between a preposition and the following noun phrase, or between the verb and its arguments (subject, object, etc.) that are frequently distant from one another.

As already said, this approach requires "parsers" (automatic syntactic analyzers) for the various languages to be processed. However, these are complex tools, far from perfect, whose quality varies considerably depending on the language considered. Applying such tools to machine translation is thus a complex operation.

One basic approach for the integration of a parser in a machine translation system is to analyze the structure of the source sentence (so as to produce a "syntactic tree") and then, for each level of the syntactic tree, try to determine an equivalent structure in the target language. As one can imagine, identifying equivalent structures between different languages is a daunting challenge that results in a lot of "silence" (in other words, many of the sentences to be translated include structures that will never have been observed up to that point in the training data; i.e., in the available aligned bilingual corpora). It is therefore necessary to imagine a process that makes it possible to translate even if the system cannot analyze the whole sentence properly from a syntactic point of view. Different strategies have been tried, especially "generalization tricks" inherited from the example-based paradigm, where one tries to find a similar structure if a sentence or a part of a sentence cannot be properly analyzed.

It also happens, for some language pairs, that a parser exists for one language (the source or the target language) but not for the other one. Some experiments have been done to integrate the parser only on one side of the translation process, with mixed results.

Globally, the idea of integrating syntax in the translation process seems, of course, promising. However, the approach is still in its infancy and, for the time being, has not obtained better results than the simpler models using direct translation techniques based on segments. The first reason is certainly the highly variable performance of syntactic parsers. Moreover, if the tool makes an error, it will "percolate" throughout the entire translation process. The "silence" issue (i.e., syntactic structures for which the parser is unable to provide a relevant analysis or for which there is no direct equivalent in the target language) is the other main cause of this limited performance. The problem is in a way logical: things are often phrased differently in different languages. It is thus not surprising that no directly equivalent structures can be found in the target language. This is surely a serious challenge for the integration of syntactic parsers in the translation process.

So far, syntax seems to be especially useful in certain specific contexts or for certain languages like German, where the verb is frequently broken into two elements as we have seen. The main benefit of the approach is that it performs a local and limited syntactic analysis to resolve some of the specific problems of each language. All of this makes syntax both a serious avenue to improve existing systems and a difficult solution to implement in practice.

Alignment Models Accounting for Semantics

Semantic analysis clearly remains a crucial perspective for the domain. We have seen that the systems thus far have mainly used large bilingual corpora as their main source of data. We have also seen that the integration of linguistic knowledge has brought very little benefit to the different systems so far. However, most experts in the field think that, despite everything, it will be necessary in the short or medium term to integrate semantic information in order to overcome current limitations.

In fact, semantic resources are already used, even by well-known available systems. For example, Google Translate integrates Wordnet, a large lexical database developed at Princeton University for English. Google also uses other semantic resources depending on the language considered. Semantic resources may be useful to disambiguate ambiguous elements: for example, if the system has to translate "*the tank was full of water*," it must decide if "*tank*" represents "a container that receives something" or "a military vehicle." Wordnet offers a long list of synonyms for "*tank*" as container, including the word "*bucket*." The sentence "*the bucket was full of water*" is well attested (whereas "*the armoured vehicle was full of water*" is not nearly as attested), which compels the user to identify "*tank*" as representing a container (or, more precisely, a "reservoir") in the sentence.

The integration of large databases of synonyms makes it possible to group certain words (essentially nouns and verbs) into semantic classes, thus providing a more abstract representation. The system can then spot identical structures beyond surface divergence. The semantic analysis may also focus on identifying specific sequences, such as named entities (proper nouns, dates, etc.), again for the purpose of providing more general and more abstract representations. This is very close to the strategies we have already seen for example-based translation (see chapter 8).

A deeper semantic analysis could provide a representation of the semantic structure of the sentences to translate. This applies particularly to the verb, its arguments, and their role in the sentence (subject, object, or, even better, agent, patient, temporal argument, etc.).

This corresponds to the vertex of Vauquois' triangle (see chapter 3). Although many research groups currently focus on these issues in natural language processing, the performance is still too low to be applied as they are to any text, as must be the case for machine translation systems. This remains an open line of research for the years to come, but it will most likely take several more years before efficient systems integrating a semantic analysis become available, given the difficulty of the tasks.

CHALLENGES AND LIMITATIONS OF STATISTICAL MACHINE TRANSLATION

Following the historical overview, it is worth examining some of the limitations of statistical machine translation systems. One fundamental question is related to the approach itself: is it really possible to translate just by putting together sequences of words extracted from very large bilingual corpora? What quality can be expected with such an approach? We will return to this topic at the end of this chapter.

In the meantime, we wish to address two other issues with this approach. It is clear that sentence alignment works better when there is a certain proximity between the language of the source text and the language of the target text. This has an impact on the performance one can expect from a machine translation system: how far can we go with the statistical approach when it is applied to genetically distant languages? Is translation from Chinese or

Arabic to English doomed to lag behind? Lastly, statistical translation presupposes the availability of large bilingual training corpora. Thus, problems arise once one leaves the very restricted circle of the best-represented languages on the Internet.

The Question of Language Diversity

Languages Distant from English

As we have seen, the majority of machine translation systems today use a statistical approach. Identifying translational equivalents at word or segment level works better when similar languages are involved, since these languages will share a similar linguistic structure. This can be seen very directly in the performance of various systems (see chapter 13): it is easier to translate between French or German and English than between Arabic and Japanese or Chinese and English. Even between these languages there are crucial differences. For example, translating from German into English works better than translating from English into German, since compound words in German (i.e., the complex combination of several simple words into a single string of characters) remain problematic for automatic processing. We know relatively well how to automatically decompose existing compounds in German; therefore, compounds are not so problematic when translating to English. It remains quite difficult, however, for an automatic system to generate correct compound words in German, which means that poor translations will be produced when translating to German.

Translating into Japanese or, more recently, Chinese or Arabic has generated a significant amount of research. Performance, compared to those obtained with Indo-European languages such as French or Spanish, remains lower, as these languages have a structure that is very different from English. For these languages, the development of hybrid systems integrating a statistical component but also advanced linguistic modules that take language specificities into account will likely be the main source of progress in the years to come. It is already the case for morphologically-rich languages, for example (i.e., languages for which a lot of different surface forms can be generated from one basic linguistic form): language-specific modules dealing with morphology issues are very helpful in enhancing the overall performance of natural language processing systems in this context.

The Case of Rare Languages and the Return of Pivot Languages

All statistical systems require a huge amount of bilingual texts in order to work satisfactorily. Corpora made of millions of aligned sentences are nowadays commonplace.

As Mercer said, "there is no data like more data" (see chapter 7).

Consequently, it is clear that beyond a handful of languages that are widely used on the Internet, the systems' performance decreases considerably, especially if one of the languages (source or target) is not English. The quantities of data available on the Internet for these languages are simply insufficient to obtain good performance. Some techniques have been developed to overcome the lack of bilingual data. For example, it is possible to obtain more information from large monolingual corpora, but this remains insufficient for the task.

A popular strategy consists in trying to design translation systems that use English as a pivot language, in order to overcome to a certain extent the lack of training data. The idea is that when there are not enough bilingual data between the two languages (for example, between Greek and Finnish), a solution is to translate first from Greek into English, and then from English into Finnish. The approach is simple and can provide interesting results in some contexts. However, it does not fully solve the problem: the quality of the translation to and from English is sometimes mediocre and applying two steps of translation instead of one also multiplies the likelihood of errors.

The problem of multiple translations is well known and can be observed even when one tries to translate iteratively between the same two languages (for example, from English to French and then back to English). The prototypical example is probably this Biblical sentence: "*The spirit is willing, but the flesh is weak.*" The story goes that this was translated into Russian and then translated back into English as "*The whiskey is strong, but the meat is rotten.*" Yet this is indeed an apocryphal example.[1] The longevity of this invented example is due to the comical nature of the resulting translation, but it also illustrates the fact that multiplying translation steps amounts to gradually straying away from the original text until an incomprehensible translation is obtained.

Despite these well-known problems with the "pivot approach," many critics have pointed out that Google Translate increasingly uses English as a pivot language. This often leads to strange results. As Frédéric Kaplan noted on his blog,[2] with Google Translate, "*Il pleut des cordes*" can be translated to Italian as "*Piove cani e gatti.*" Likewise, "*Cette fille est jolie*" strangely enough transforms into "*Questa ragazza è abbastanza*" ("*this girl is quite*"). These major errors are due to the fact that English is used as a pivot language. In the French expression, "*Il pleut des cordes,*" Google identifies a frozen expression that should not be translated word for word. "*It rains cats and dogs*" is thus a good English translation in this context, but the system then fails to find the equivalent expression in Italian and then just performs a word-for-word translation, which is poetic but not really accurate. As for *"joli"* in the other example, the system identifies *"pretty"* as an equivalent adjective but then seems to confuse the adjectival and the adverbial values of *"pretty"* in English, hence the translation related to *"quite"* (*"abbastanza"*) instead of *"pretty"* as *"nice"* or *"beautiful."*

These examples quickly lose their relevance as Google constantly updates its system: Kaplan's blog post dates from November 15, 2014, and on December 1, the translation *"Il pleut des cordes"* in Italian had already become *"Piove a dirotto,"* which is a correct translation. It is clear that Google works intensively to improve the translation of frozen and multiword expressions like *"il pleut des cordes"* in French or *"it rains cats and dogs"* in English: these expressions are special since they should be translated as a whole and not literally. There are obviously no cats or dogs falling from the sky, when one says the English expression! This is a major source of enhancement for current systems, but the task is huge since there are a large number of frozen expressions in any language.

Clearly, using English as a pivot language has several implications. It reinforces the dominant position of English as a world language and its cultural hegemony. Besides this, and despite official discourse promoting language diversity, it is clear that the research effort is mainly directed toward dominant languages (the handful of languages dominating the Internet). Google Translate can officially

Despite official discourse promoting language diversity, it is clear that the research effort is mainly directed toward dominant languages (the handful of languages dominating the Internet).

translate between more then 100 languages, but in practice the results are very uneven and almost unusable for some languages.

Finally, we must keep in mind that the advances since the 1990s in the field of statistical machine translation are related to the fact that more and more data are available, but also to an increase in computing power. The deep learning approach uses some algorithms that were to a certain extent described in the 1980s but that researchers were unable to apply in practice due to the limitations of machines at the time. Things have changed, and it comes as no surprise that it is the companies with huge computational capacities that are now developing the most efficient systems.

How to Quickly Develop Machine Translation Systems for New Languages?

Despite the current predominance of statistical machine translation systems, we should note the persistence of rule-based systems. At the same time, the majority of historical systems are now said to be "hybrid": they attempt to combine the benefits of a symbolic approach (i.e., dictionaries with very wide coverage, transfer rules between languages) with the recent benefits of statistical techniques. Lastly, for rare languages with too few data to make it possible to develop statistical systems, rule-based systems remain the norm.

Hybrid Machine Translation Systems

Following the success of statistical translation systems, the majority of traditional systems (based on large lexicons and transfer rules) gradually tried to incorporate statistical information in their approach. A striking example is Systran, which was the proponent of the rule-based approach in the early 2000s before developing a hybrid approach to translation, based on both knowledge and statistics. There is, in particular, a clear benefit in using a language model to control the fluency of the translation generated: Systran first used statistics to correct the output and produce more fluent translations.

In practice, statistical information can be integrated in an infinite number of ways into systems that otherwise manipulate symbolic information. It is, for example, possible to design modules that will dynamically adapt the system (the dictionaries and the rules used for translation) according to the domain, for example in the medical, legal, and information technology domains. The statistical approach can also help choose the correct translation at word level. For example, if the system detects a text related to the military domain, the meaning of "*tank*" as a military vehicle will be preferred to its other meaning of receptacle (for more on this subject, see chapter 10).

The overall idea is of course to combine the richness of existing resources, which are sometimes the result of years of research and development, with the efficiency of statistical approaches. It is now possible to say that the gap has largely closed between the rule-based and the statistical approaches: today most commercial systems are hybrid. We saw above that even Google is integrating more and more semantic resources into its system, making it a truly hybrid system.

The Survival of Rule-Based Systems

Finally, the survival of conventional systems based on rules and bilingual dictionaries only (with little or no statistics at all) should also be noted. When there are too few bilingual corpora available, the statistical approach is no longer of great interest.

Different systems exist for the development of rule-based systems. Apertium (www.apertium.org) is one such platform. The system was originally intended to be used for closely related languages that required a limited number of transfer rules (for example, for two dialects or two closely related languages differing mainly in their vocabulary but not so much in their syntax). But after a while the platform turned out to be also useful for all underrepresented languages on the Internet. It has even since specialized in processing rare languages like Basque, Breton, or Northern Sami (a language from the north of

Scandinavia), which are available in this system (the system has data available for about 30 languages, but translation is only possible for 40 translation pairs, most of the time in one direction only). The performance is variable depending on the language pair considered, and most of the implemented language pairs are based on bilingual dictionaries and reordering rules. One of the goals of this project is to promote rare languages and provide access to texts that would not be translated otherwise. It also aims to generate interest for these languages, which for the most part are endangered.

A Current Challenge: The Rapid Development of Translation Systems for New Language Pairs

Here we must say a few words on one of the current challenges in the field of machine translation: the rapid development of translation systems for languages that have not been covered up to now. This concerns mainly the defense and intelligence industry; surveillance and intelligence needs evolve rapidly depending on geopolitical risks (see chapter 14, dedicated to the machine translation market).

From a technical point of view, the challenge is to collect bilingual corpora very quickly for the languages considered. While automatic corpus collection is a well-known technique nowadays (see chapter 7), the volume of data collected is often insufficient in practice to develop operational machine translation systems. Since most of the time the target is a language that is distant from English, the result is not as good as for closely related language pairs.

In this context, the statistical approach is not the predominant one. The task consists for the most part in developing large bilingual dictionaries manually. Monolingual corpora are collected and processed so as to automatically produce a large list of words in the target language. Then an automatic process or, more likely, a team of linguists will proceed to provide word translations so as to make it possible to quickly develop a rudimentary translation system. The production speed of such a system closely depends on the number of linguists that can be hired for the project, but this type of demand currently represents a significant part of the machine translation business.

Too Many Statistics?

The current situation of machine translation poses many fundamental questions for the field. Is semantics necessary to translate or can we settle for statistics? Also, can we say that current systems, specifically those based on statistics, are completely stripped of semantics? Finally, are the approaches used today going to allow for significant progress or, on the contrary, can we anticipate some strong limitations that will prevent improvements in the near future?

Primary Limitations of Statistics-Based Systems

Experts in the field have discussed all these questions. Systems are now based on highly technical machine-learning approaches, and linguistics has been left aside. As we will see in chapter 14, commercial issues are important and put pressure on developers to find efficient short-term solutions. At the same time, the progress of systems is measured annually in evaluation conferences: competition is strong, leaving little time for reflecting on the current state of affairs.

It is first of all crucial to recall that since the 1990s we have seen real and undeniable progress in the field of machine translation. Statistical methods made it possible to better process a large number of frequent and important phenomena (such as the search for the best translation at word level, the management of local ambiguities, the relative contribution of different linguistic constraints when they seem to contradict one another, etc.). These phenomena were generally not solved satisfactorily by rule-based methods. However, the success of statistical models was such that even some of its proponents have called recent progress into question.[3] Some local phenomena are relatively satisfactorily solved by current statistical techniques, but more complex phenomena should also be taken into account.

Many very frequent linguistic phenomena (agreement, coordination, pronoun resolution) would indeed require a more complex analysis. They are poorly or not at all addressed by statistical systems (but note that this also applies to most rule-based systems that also focus on local context). The state of the art is just too limited to deal with these complex issues. Syntactic analysis is hard, but semantics is even harder and we are still far from knowing how to address this kind of problem properly.

Statistics Do Not Exclude Semantics

One interesting question concerning recent approaches to machine translation is the status of statistics in this context. It is widely assumed that statistics are opposed to semantics: on the one hand calculation, on the other hand representation of word and sentence meaning. Yet this opposition is too crude. As we have seen in the previous chapters, statistics make it possible to accurately model the different meanings of words according to the context. Statistics are also efficient to find translational equivalences at word or phrase level.

This raises another question: what, by the way, is the meaning of a word? How can we represent it? It is indeed very difficult to define the meaning of words precisely. This is the job of lexicographers who spend years developing dictionaries, but the enumeration of word meanings they propose is known to be quite subjective and does not always correspond with word usage in context. Moreover, definitions vary significantly from one dictionary to the other, especially for abstract concepts and functional words.

Given this state of affairs, it should be taken into account that although accurate definitions are hard to find, it is easy for any speaker of a language to give synonyms of a given word or examples of word usage in context. The various word meanings correspond, in fact, to various contexts of use. The challenge thus lies in defining and characterizing the notion of "context." In other words: how can one determine the various meanings of a given word just by observing its usage within a very large corpus? How can usage patterns be identified? Lexicographers (responsible for writing dictionaries) generally use a multitude of tools and criteria to define the various word meanings, and they try to be comprehensive, regular, and coherent. Statistics help automate the process and obtain results that are often different but always interesting.

Multilingual corpora provide a direct and quite natural model for the question of word meaning. The more vague or ambiguous a word is, the more it will match with a variety of different words in the target language. In contrast, the more stable and fixed an expression is (for example, "*cryptographie*") the more it will be aligned with a limited number of words (such as "cryptography") because the word is not (or less) ambiguous. For the same reasons, the approach is able to recognize the fact that a multiword expression (like "*pomme de terre*" in French) corresponds in the target language to a single word ("*potato*" in English). The same is true for frozen expressions ("*kick the bucket*" or "*passer l'arme à gauche*," which both mean "*to die*"). Statistical approaches may seem too simple or too crude, but the system will not produce "*frapper le seau*" for "*kick the bucket*" or "*pass the weapon to the left*" for "*passer l'arme à gauche*" if it has been properly trained (but these kinds of problem may occur when frozen expressions are not properly recognized, as already said in this chapter). These examples show that statistical analysis therefore leads to a direct modeling of polysemy, idioms, and frozen expressions, without any predefined linguistic theory.

It can even be claimed that the type of representation obtained from a statistical analysis is more appropriate and cognitively more plausible than what formal approaches propose. Notions such as ambiguity and polysemy (or, in other words, meaning) are closely linked to usage and are not absolute notions. In this respect, it is understandable that statistical analysis can help define the various meanings of a word, the different contexts in which it appears, and so on. Numerous linguists and philosophers have defended such ideas, from Ludwig Wittgenstein to John Rupert Firth, the latter of whom is the author of the famous quote: "You shall know a word by the company it keeps" (i.e., you know the meaning of a word by its context of use). This remark has been cited again and again in modern natural language processing texts. Current approaches may not have anything to say about Wittgenstein or Firth, but they are beyond doubt very close to the text, and they have eliminated everything that was "metaphysical" in semantics (the quest for artificial modes of representation, the idea of a universal language, the goal of transforming sentences into logical forms, etc.). It is possible that current approaches will form the basis of a new theory of meaning.

However, the remarks made at the beginning of this chapter should be kept in mind: representations used by statistical machine translation systems remain local, for the most part, which does not make it possible to address many of the fundamental problems related to semantics. Lexical semantics (i.e., the meaning of words) is relatively well formalized today, but propositional semantics (i.e., the meaning of sentences and the relations between them) remains very difficult to achieve and thus, to a large extent, a "terra incognita." This is what a new approach known as deep learning, or neural machine translation, which we describe in the following chapter, is trying to solve.

DEEP LEARNING MACHINE TRANSLATION

Over the past several years, a new type of statistical learning called "deep learning" or "hierarchical learning" has emerged in the wake of neural networks. Neural networks were originally inspired by the biological brain: neurons transmit and process basic information, from which the brain is able to build complex concepts and ideas. Artificial neural networks, like the brain, are supposed to be able to build complex concepts from different pieces of information assembled in a hierarchical manner. But, as outlined in Goodfellow et al. (2016, p. 13): "the modern term 'deep learning' goes beyond the neuroscientific perspective on the current breed of machine learning models. It appeals to a more general principle of learning multiple levels of composition, which can be applied in machine learning frameworks that are not necessarily neurally inspired."

This approach has received extensive press coverage. This was particularly the case in March 2016, when Google Deepmind's system AlphaGo—based on deep learning—beat the world champion in the game of Go. This approach is especially efficient in complex environments such as Go, where it is impossible to systematically explore all the possible combinations due to combinatorial explosion (i.e., there are very quickly too many possibilities to be able to explore all of them systematically).

The complexity of human languages is somewhat different: the overall meaning of a sentence or of a text is based on ambiguous words, with no clear-cut boundaries between word senses, and all in relation to one another. Moreover, word senses do not directly correspond across different languages, and the same notion can be expressed by a single word or by a group of words, depending on the context and language considered. This explains why it is impossible to manually specify all the information that would be necessary for an automatic machine translation system, but also why the translation task has remained highly challenging and computationally expensive up to the present time. In this context, deep learning provides an interesting approach that seems especially fitted for the challenges involved in improving human language processing.

An Overview of Deep Learning for Machine Translation

Deep learning achieved its first success in image recognition. Rather than using a group of predefined characteristics, deep learning generally operates from a very large set of examples (hundreds of thousands of images of faces, for example) to automatically extract the most relevant characteristics (called *features* in machine learning). Learning is hierarchical, since it starts with basic elements (pixels in the case of an image, characters or words in the case of a language) in order to identify more complex structures (segments or lines in an image; sequences of words or phrases in the case of a language) until it obtains an overall analysis of the object to be analyzed (a form, a sentence). An analogy is often drawn with human perception: on the one hand, the brain analyzes groups of simple items very rapidly in order to identify higher-level characteristics, and on the other hand, it recognizes complex forms from characteristic features, and can even extrapolate a complex representation from partial information (this is essentially what happens in the case of the Necker cube, where the brain infers a three-dimensional representation from a two-dimensional drawing; see figure 1 in chapter 2).

In the case of machine translation, deep learning makes it possible to envision systems where very few elements are specified manually, the idea being to let the system

In the case of machine translation, deep learning makes it possible to envision systems where very few elements are specified manually, the idea being to let the system infer by itself the best representation from the data.

infer by itself the best representation from the data. This was, in a way, already the idea with purely statistical models, but we have seen that in fact many parameters had to be adjusted manually. For example, five models were proposed in the early 1990s by IBM for machine translation, each model introducing a different manually defined representation to correct certain defects of the previous model. Deep learning, on the contrary, makes it possible, at least in theory, to learn complex characteristics fully autonomously and gradually from the data, without any prior human effort.

A translation system based solely on deep learning (aka "deep learning machine translation" or "neural machine translation") thus simply consists of an "encoder" (the part of the system that analyzes the training data) and a "decoder" (the part of the system that automatically produces a translation from a given sentence, based on the data analyzed by the encoder). We have already seen this vocabulary for the traditional statistical approach (see chapter 9), but here the encoder and the decoder are based uniquely on a neural network, whereas traditional statistical approaches use a combination of modules (typically, a language model and a translation model for the encoder part) to be able to use different optimization strategies. In a neural network, each word is encoded through a vector of numbers and all the word vectors are gradually combined to provide a representation of the whole sentence.

In a way, we can say that deep learning machine translation adopts a more traditional architecture than statistical machine translation, since the encoder can be seen as the analyzer of the source language, whereas the decoder generates the translation in the target language (as in Vauquois' triangle; see figure 2 in chapter 3).

With deep learning, the simultaneous management of various types of information enables more reliable decision making. These models are said to be hierarchical, but they are in fact multidimensional, meaning that each element (word, phrase, etc.) is placed within a richer context. Following the famous motto "you shall know a word by the company it keeps" (from the British linguist Firth), the approach is based on the hypothesis that words appearing in similar contexts may have a similar meaning. The system thus tries to identify and group words appearing in similar translational contexts in what is called "word embeddings." This approach makes the process a lot more general and thus more robust than what we have seen so far: it may not be a problem if a word is rare, since other words appearing in similar contexts may indicate a valuable translation. The fact that a word has different meanings is not a problem either, since it can belong to different embeddings, reflecting different contexts of use.

A second characteristic of deep learning approaches is that these models are said to be "continuous." It was already partially the case with statistical machine translation, since in this framework words can be considered "more or less" similar to each other (meaning that all pairs of words have a similarity score between 0 and 1). This representation seems cognitively more plausible than the one given, for example, by traditional synonym dictionaries: there are indeed plenty of cases where words have a "more or less" strong similarity without being strictly synonyms. The deep learning approach generalizes this idea, so that words, but also higher linguistic units, like phrases, sentences, or simply groups of words, can be compared in a continuous space, which makes the approach highly flexible and able to identify, for example, paraphrases.

Lastly, it should be noted that closely related words inside a sentence are also gradually identified and grouped together during the analysis. This is why the deep learning approach is said to be hierarchical, since it is able to discover structure (relations between words or groups of words) inside a sentence, based on regularities observed in the thousands of examples given to the system during training: although deep learning systems do not directly encode syntax, they are supposed to be able to automatically identify relevant syntactic relations.

In brief, rather than having different modules considering different parts of the problem at a time, the deep learning approach to machine translation considers directly the whole sentence without having to decompose it into smaller segments, and also considers all kinds of relations in context at the same time. The fact that these relations can be vertical (groups of similar words that can fill a position in a sentence) or horizontal (syntactically related groups of words in a sentence) makes the approach highly flexible and cognitively interesting, but also computationally challenging.

There have been in fact several generations of artificial neural networks (the approach has only recently become called "deep learning"). Neural networks were actually invented in the 1950s and were blooming again in the 1980s—but the computational power of the machines at the time did not allow for managing the complexity of the representations involved (Goodfellow et al., 2016, pp. 13–28). Even today, the training phase of such a system may last for days. Specific processors and programming techniques (GPU-accelerated programming) are used to speed up the process. The approach is also, in reality, a lot more complex and abstract than what we have just described. A context is, for example, encoded through a vector of numbers, each number representing a feature (an abstract property automatically discovered by the neural network from the regularities in the corpus), with the length of the vector corresponding to a predefined value. A recent evolution consists in adjusting dynamically this value so that more or less information can be encoded depending on the complexity of the task.

It should also be noted that the approach remains empirical, especially when it comes to defining the architecture of the neural network used (e.g., the number of layers in the neural network, the length of the vectors used) as well as other parameters (e.g., the way unknown words are processed); there is little theoretical basis for these choices, which are mainly based on system performance and efficiency. These systems are sometimes criticized as lacking theoretical foundation for this reason.

Nonetheless, deep learning is a real step forward and has enabled significant improvements in the field of image recognition, speech processing, and, more recently, natural language processing. Some researchers today are going so far as to challenge traditional disciplines such as syntax, because through deep learning it is possible to infer structure from the data. In other words, it would be better to let the system determine on its own the best representation for a given sentence![1] It remains, however, necessary to put these claims into perspective: probably because of the dramatic amount of variation in sentences, systems still frequently fail to recognize the overall sentence structure, which can lead to major translation errors. Still, deep learning opens a window toward the resolution of such problems, hence the great success of this technique among researchers in the domain.

Current Challenges for Deep Learning Machine Translation

Until recently, machine translation systems based on deep learning performed well on simple sentences but still lagged behind traditional statistical systems for more complex sentences. There were different reasons for that, as explained by the Google team working on the question: first, training neural networks for the task is still difficult due to their complexity, especially the number of parameters that have to be automatically adjusted. This led to various efficiency problems. Second, unknown words (i.e., words not included in the training data) are generally not processed accurately (or are just ignored) in this approach. Finally, groups of words are sometimes not translated, leading to strange and inaccurate translations. For some time, this prevented purely neural approaches from being effectively deployed in commercial systems. This is, however, no longer the case, since efficient solutions are emerging.

Optimization techniques have been used to reduce learning complexity in the encoder as well as the decoder. What are called "attention" mechanisms also play a growing role in neural network architectures, especially for machine translation. The "attention module" helps create connections between the encoder and the decoder, a bit like the way in which a transfer rule in a rule-based machine translation system formalizes how a linguistic structure in the source language must be rendered in the target language. However, the analogy should not be taken too far: here, again, the process is a lot more abstract than what can be found in traditional transfer rules.

Intuitively, the approach is based on the fact that some words in the source sentence are especially important when it comes to translating a specific word in the target language (or, put differently, not all the words in the source sentence are equally relevant at any time in the translation process). When one translates from French to English, both languages have relatively similar structures, so that the translation process can be relatively sequential, especially when dealing with short sentences (10 words or less): knowing the n previous words in the source sentence is often enough to produce the next word in the target language. Longer sentences have a more variable word order; attention mechanisms then help the system to dynamically focus at any time on the most relevant parts of the sentence to be translated. It can be useful, for example, to keep in memory the fact that there is a link between a verb and its subject (especially if a long sequence of words is inserted between them): this link may play a prominent role, for example, to control agreement when the system generates the verb in the target language. It has been shown that attention mechanisms capable of focusing on relevant source words considerably increase overall translation quality. Attention mechanisms are especially useful to deal with long sentences, but it is also assumed that they will play an even stronger role when dealing with typologically diverse languages, for example when translating between English and Japanese, since in Japanese the verb is located at the end of the sentence.

The unknown word problem is a real challenge for deep learning approaches, since only words contained in the training data are part of the model and can thus be translated. When statistical systems were modular, it was easy to integrate a module dealing with unknown words. It is more difficult with the deep learning model, which offers a more holistic approach. Some "patches" have, however, recently been found. Because unknown words are in fact often proper nouns or numbers, some systems just "copy" the unknown word from the source to the target language. When writing systems are different (for example, when translating from Arabic or Chinese to English), transliteration works reasonably well and can be a valuable solution. Unfortunately, it is also quite frequent to find unknown words that are neither proper names nor numbers. A working solution consists then in trying to decompose unknown words into smaller units so as to find relevant cues to help the translation process, but this approach is not fully satisfactory;

unknown words remain an open problem for deep learning approaches.

Lastly, it is necessary to verify that the system translates the whole sentence and does not omit sequences of words from the source sentence. This may seem surprising for such sophisticated machine translation systems, but the truth is that since the sentence is analyzed globally and not decomposed into segments, as in previous statistical approaches, the system can fail to translate some words or phrases because they are loosely related to the core of the sentence or for other more mysterious reasons. To solve the problem, the Google team proposes to implement a length penalty to help the system favor the longest translations, so as to decrease the weight of a candidate translation in which part of the initial sentence is not translated. This trick is simple and efficient, but this problem shows that it is hard to understand and analyze the way a neural system works, since the internal representation of the data is purely numerical, huge and complex, and more importantly not directly readable by a human being. A very promising trend of research is however to try to get meaningful representations of the internal model calculated by the neural network, so as to better understand how the whole approach works.

The deep learning approach to machine translation (or neural machine translation) has proven efficient, first, on short sentences in closely related languages, and more recently on long sentences as well as more diverse languages. Progress is very quick, and the deep learning approach can be considered a revolution for the domain, as was the statistical approach at the beginning of the 1990s.

It is interesting to note that deep learning approaches spread very quickly. All the major players in the domain (Google, Bing, Facebook, Systran, etc.) are moving forward to deep learning, and 2016 saw the deployment of the first online systems based on this approach. This can be contrasted with the advent of the statistical approach, which took several years to dominate the market and supersede rule-based systems. The deployment of deep learning solutions is much faster. It also means that the approach is now robust and mature enough to outperform statistical approaches.

However, this approach is still in its infancy, and rapid progress can be expected in the near future. More efficient solutions will be proposed to the problems described above, for example to deal with unknown words. It should also be noted that some actors of the domain still favor a more modular solution so that specific issues can be solved more accurately (and neural networks are then just introduced locally, in some modules of a traditional statistical machine translation system, for example). In a way, this may go against the philosophy of the neural approach, since processing data at sentence level is the source of most of the improvements described in this chapter. The future will tell which approach is the best.

As a conclusion of this chapter, we should remember that the world chess champion was beaten by a computer in 1997, the world Go champion was beaten by a computer in 2016, but no computer is able to translate accurately between two languages even today! This shows the complexity of natural languages.

THE EVALUATION OF MACHINE
TRANSLATION SYSTEMS

As we have seen, translation systems have been the subject of intensive research since the renewal of the field during the 1990s following IBM's experiments, described in chapter 9. The development of the web drove the main Internet companies to look into the problem, which also helped revive research. The question then arose of how to measure the quality of the systems. How can two systems be compared? How can the development of a single system over time be measured and its improvement tracked?

Additionally, we saw in chapter 2 the difficulty in defining what makes a good translation. It is thus clearly difficult to evaluate the quality of a translation, since any evaluation involves some degree of subjectivity and strongly depends on the needs and point of view of the user. The IBM team, in the seminal 1988 article (see chapter 9), quickly raised the issue by mentioning literary

It is clearly difficult to evaluate the quality of a translation, since any evaluation involves some degree of subjectivity and strongly depends on the needs and point of view of the user.

translation. The last word in Proust's *In Search of Lost Time* is the same as the first word of the first volume (the novel actually begins with "*longtemps*" and ends with "*temps*"). Literary translators must focus on these types of details, which are fundamental for the interpretation of a work, but IBM immediately dismissed the problem by making it clear that machine translation has nothing to do with literary translation. Therefore the IBM team did not address these kinds of details, which exceed the scope of current research.

Despite the difficulties we have mentioned, it appeared necessary to devise some evaluation methods that are reliable, quick, reproducible, and if possible inexpensive. To do this, specific evaluation datasets were produced and evaluation campaigns were organized.

The First Evaluation Campaigns

Since the beginnings of machine translation, evaluation has been perceived as necessary, more so than in other fields of natural language processing, probably because machine translation was seen from the beginning as an applicative field and very concrete results were expected. We have seen in this regard that the ALPAC report was very negative and rather skeptical about the quality that could be hoped for from such systems (see chapter 6).

At the beginning of the 1990s, with the renewal of research based on a statistical approach originally proposed by IBM, the need to measure machine translation systems was again felt. As is often the case in the field of natural language processing, it was an American funding agency, the Advanced Research Project Agency (ARPA, later known as DARPA[1]), that initiated research in this area. A 1994 article (White et al., 1994) reviewed the first attempts at evaluation from the beginnings of research on machine translation. The article specifically reported the various possible strategies and their limits, described below.

Comprehension Evaluation   To assess comprehension, professional human translators first translated English newspaper articles into different languages. Machine translation systems then translated the text back into English, and human analysts answered "multiple choice questions about the content of the articles" to evaluate the automatic translations, as explained by White and his colleagues. The number of questions the reader of the translation was capable of answering correctly determined the quality of a system. Because the first campaigns focused on a limited number of systems capable of translating into English from different languages, this method was well suited to the task: the text to be translated was provided in various languages, and the translations into English could then be compared. This test was initially named "direct comparability" because it was supposed to allow for a direct comparison of different systems from different source languages.

White et al.'s review of this kind of evaluation was rather mixed. The translations provided by the human translators, although they were supposed to be translations of the same text, were in fact all different and may have posed specific problems for a machine translation system. It was therefore difficult to know if the comprehension errors were to be attributed to the way the original text was phrased or to the translation system itself (not to mention the potential problems related to the interpretation of the text by the reader in charge of the evaluation). The method was eventually abandoned as a means of overall evaluation, but was, however, kept for evaluating the "informativeness" of the text with regard to the original text.

Evaluation Panel   The most obvious way to evaluate the quality of translations is to appeal to human judgment, notwithstanding the great degree of subjectivity inherent in human judgment. DARPA resorted to this in the early 1990s: the judges had to evaluate the quality of the translation produced, taking into consideration the lexical, grammatical, semantic, and stylistic aspects of the translated texts. As White and his colleagues pointed out, this method seemed attractive in that it also served to evaluate the quality of human translations.

However, this type of evaluation encountered two major difficulties. First, from a practical point of view, it was very difficult and expensive to bring together a group of experts for the entire duration of an evaluation campaign. More importantly, the types of errors in the texts produced automatically were so diverse that it was extremely difficult for an expert to assign an overall score to a text (in practice, this score varied enormously from one expert to another, depending on the importance attached to such or such a type of error by a given expert). Despite various attempts to homogenize notation strategies, wide variations between the scores assigned by experts remained. This evaluation method was thus not judged fully satisfactory, and the quality panel evaluation method was abandoned.

Adequacy and Fluency After the previous attempts involving human experts, DARPA then resorted to two evaluation scores: adequacy and fluency. As White and colleagues described of this machine translation (MT) evaluation method: "In an adequacy evaluation, literate, monolingual English speakers make judgments determining the degree to which the information in a professional translation can be found in an MT (or control) output of the same text." These pieces of information were generally fragments "containing sufficient information to permit the location of the same information in the MT output." The fluency score aimed to verify correct sentence formation, the task being "to determine whether each sentence is well formed and fluent in context." These criteria proved easier to use than those previously mentioned and became the standard set of methodologies for the DARPA MT evaluation. However, these measures remained subjective, and it has been shown that the scores assigned by experts still varied significantly.

Human-Assisted Translation   A final evaluation strategy takes as a starting point the fact that no automatic system is capable of producing a perfect translation. It therefore seems relevant to evaluate to what extent a machine translation can help a human translator obtain a good translation. The experiments conducted in the early 1990s involved a novice human translator, who was supposed to derive greater benefit from an imperfect translation than an experienced human translator (who was assumed to be more capable of seeing how to "properly" translate a sentence without the help of an automated process). The evaluation focused on the comparison between the results of the automatic process with the results of the improved translation done by the translator.

White et al. (1994) reported that this type of evaluation seemed to give interesting results. Nonetheless, several factors made it very difficult to use in practice. First, it was very difficult to control for the "beginner" status of the human translator. There is a great deal of variation from one individual to the next, which makes any comparison very subjective. Secondly, the added value of the various components of the automated system (especially the components managing the interaction with the translator, which were not directly part of the machine translation system) was difficult to assess. Finally, the majority of the automated systems evaluated already included modules that required some kind of interaction with the user, which made the result of the purely automated translation system difficult to isolate. [2]

In the mid-1990s, three measures were mainly used for evaluation: comprehension, adequacy, and fluency of the generated text. These three measures were interesting but relied largely on human judgment, which is known to be costly and partially inconsistent. This led experts in the field, toward the end of the 1990s, to try to find entirely automatic measures without human intervention.

Looking for Automatic Measures

Automatic evaluation measures aim at answering a simple question: given one (or several) reference translation(s), how can the quality of an automatic translation be measured? Similar questions arose around the same time—toward the end of the 1990s and the beginning of the 2000s up to the present day—in regard to automatic summarization, for example. While this question seems simple, finding an answer is obviously much more complicated. Several measures were defined; we briefly present the main ones below, without going into further mathematical details.

BLEU

The principles of the Bilingual Evaluation Understudy, or BLEU, score (Papineni et al., 2002) are relatively simple. The idea is to compare a reference translation T_{Ref} with an automatically produced translation T_{Auto}. The BLEU score is calculated by truncating T_{Ref} and T_{Auto} into segments of length 1 to n, called n-grams (it is generally assumed that the most reliable result is obtained when $n = 4$) and by comparing the number of segments shared between T_{Ref} and T_{Auto}. The BLEU formula also includes a parameter that takes into account the length of the sentences in the automatically produced translation, so as to not favor systems producing too-short sentences.

If two texts, T_{Ref} and T_{Auto} are identical, then the BLEU score is 1 (all segments from T_{Auto} are also part of T_{Ref}). If no segment is shared, the score is 0. In other words, the closer T_{Auto} is to the reference translation, the greater the number of shared segments will be, and the closer the BLEU score will be to 1. To improve the robustness of the result, it is possible to compare T_{Auto} with several reference translations (T_{Ref}) without changing the overall idea.

NIST

The National Institute of Standards and Technology (NIST), an American organization organizing evaluation campaigns in various fields, developed the NIST score during the same period as the BLEU score (Doddington, 2002). It is based on the same principles: the two texts to be compared, T_{Ref} and T_{Auto}, are truncated into segments (n-grams), and the measure is based on the number of segments from T_{Auto} that also feature in T_{Ref}.

The main difference is the inclusion of an informativeness factor: the rarer a segment is, the higher its weight becomes. The NIST score is generally correlated with the BLEU score, which is logical given their broad similarity. The NIST score is meant to take better account of the informational diversity in the texts to be translated.

METEOR

The METEOR score ("metric for evaluation of translation with explicit ordering"; Banerjee and Lavie, 2005) was developed more recently and tries to better account for semantics. METEOR is based on the identification of semantically "full" words (essentially nouns, verbs, and adjectives) shared between the text to be evaluated and the reference. The idea is then to identify longer sequences of text around these full words that are shared between the two texts. As with other measures, the greater the number of shared segments, and the longer these segments are, the closer the METEOR score will be to 1.

The search for similar segments is not always based on surface forms. Words can be replaced by their stem or their lemma (changing "*running*" to "*run*") or even by synonyms if a semantic resource (such as Wordnet) is used. This makes the method more reliable and more robust, but requires adequate semantic resources, which may not be available for all languages. This is the reason why the packages implementing this measure are generally provided with a list of "supported" languages (languages for which such a resource is available).

The authors of METEOR report results that are better correlated to human evaluations than the BLEU or NIST scores. However, METEOR is more difficult to operate than other scores, and the different options (e.g., whether a given linguistic resource has been used or not during the evaluation) make the results more difficult to interpret and compare over time. METEOR is therefore less frequently used than the other scores, especially BLEU, which remains the most widely used measure despite its limitations.

Comments on Automatic Evaluation Measures

All the measures presented here rely on the comparison of short sequences of words (n-grams, with n varying generally from one to four words) between a reference text and an automatically produced translation. We have seen that some measures try to take into account richer information (lemma, synonyms) but most of the time the evaluation is just based on surface forms (i.e., words as they appear in the text). The reader may be surprised by the poverty of the information used for evaluation, which completely eliminates notions such as style, fluency, or even the grammaticality of sentences. Since evaluation simply takes into account short sequences of words, it is clear that a completely illegible text consisting of random and meaningless sentences could obtain a rather good score, provided that the sentences are made of sequences of four words shared with the text used as a reference.

These biases are well known, but are not as problematic as it may seem at first sight. Since the target is to develop operational systems, there is no incentive to pursue a system that only seeks to obtain good results without worrying about the quality of the text produced. In evaluation campaigns, the output of the system is made public, so a team that obtained good results with semantically preposterous texts would not gain any benefit from them.

More fundamentally, the gap between the complexity of the translation task and the relative simplicity of the automated evaluation methods reveals that evaluation is a real issue. It is difficult to formalize notions such as that of "a good translation," since nobody knows how to define this, let alone notions such as coherence or style. The methods used for evaluation are thus rather poor but obtain decent degrees of correlation with evaluations performed by human experts, which is considered the crucial factor for the task.

The Proliferation of Evaluation Campaigns

While machine translation was a moderately active research field during the 1980s, the renewal that took place following IBM's work during the 1990s contributed to the increased number of evaluation campaigns. Since the early 2000s, several campaigns have been organized each year throughout the world.

DARPA has organized evaluation campaigns from Chinese and Arabic into English since 2001. The texts used for evaluation are stories from news agencies. Each year since 2005, the Workshop on Machine Translation (WMT) conference has also organized an evaluation campaign concerning certain European languages; for example, in 2013, the evaluation focused on the following language pairs: French-English, Spanish-English, German-English, Czech-English and Russian-English). For each pair, both directions of translation are evaluated (for example, French to English and English to French). The organizers provide participants with an initial collection of texts for development (generally a collection of aligned sentences that participants can use to adapt their system to the task), but participants can also use their own data (other corpora, bilingual or unilingual dictionaries, etc.). Upon submitting their results, participants must say whether or not they have simply used the provided data for the evaluation or if they also used other resources.

The European Commission has strongly supported the WMT evaluation campaigns from the beginning. The WMT is largely based on the availability of the Europarl corpus, which contains the transcriptions of the European parliament debates. The corpus, available in 21 languages, is specifically dedicated to machine translation: texts are aligned semiautomatically with great precision. It is an incomparable learning corpus for the development of automatic systems (see chapter 7).

It should be noted that the WMT campaign is not interested only in the evaluation of systems. A task is specifically dedicated to evaluation measures; the quest for automatic measures more closely correlated with manual evaluation remains a research area. More recently, the evaluation of the textual quality of the automatically generated translations also appeared to be a major concern, since traditional evaluation methods based only on small fragments of texts ("n-grams") leave completely open the question of the quality of the produced text, as well as its readability.

Lessons Learned from Automatic Evaluation

Automatic evaluation is important to measure the performance and evolution of systems over time. Even if automatic measures are not completely satisfactory, they make it possible to measure evolutions that generally correlate with the perception a human has of the overall quality of the systems. In other words, it has been shown that a system obtaining scores that improve over time produces translations that indeed seem to improve according to human experts. The great differences observed in the results obtained when translating among different language pairs should also be addressed. Several features may affect performance: for example, a limited amount of training data, morphologically-rich languages that are known to be harder to process automatically, or translations between genetically distant languages.

Measuring the Difficulty of the Task According to Language Pairs

Figure 19 (Koehn et al., 2009) shows the result of an experiment on 22 European languages using the same basic translation system (Moses) and comparable training data for each language pair. The training data were the JRC-Acquis corpus, which consists of texts translated and aligned between 22 European languages. The kind of text and the quantity of data were therefore the same for each language taken into account for this experiment. The scores displayed in figure 19 are BLEU scores.

The scores are not significant by themselves, but their comparison is particularly instructive, since they reveal the difficulty of the translation task depending upon the languages under consideration. While some language pairs would no doubt obtain better results if language specificities were taken into account, the goal of the experiment was precisely to emphasize differences between languages (through evaluation scores) when using a standard translation algorithm, such as the traditional IBM model, without language-dependent optimization.

The results obtained are interesting. For example, they show the difficulty of processing languages that are very distant from English (for example, Finnish, Hungarian, and Estonian all obtained poor scores, just like Maltese). A more thorough observation of the results shows that morphologically-rich languages are more difficult to

	en	bg	de	cs	da	el	es	et	fi	fr	hu	it	lt	lv	mt	nl	pl	pt	ro	sk	sl	sv
en	–	40.5	46.8	52.6	50.0	41.0	55.2	34.8	38.6	50.1	37.2	50.4	39.6	43.4	39.8	52.3	49.2	55.0	49.0	44.7	50.7	52.0
bg	61.3	–	38.7	39.4	39.6	34.5	46.9	25.5	26.7	42.4	22.0	43.5	29.3	29.1	25.9	44.9	35.1	45.9	36.8	34.1	34.1	39.9
de	53.6	26.3	–	35.4	43.1	32.8	47.1	26.7	29.5	39.4	27.6	42.7	27.6	30.3	19.8	50.2	30.2	44.1	30.7	29.4	31.4	41.2
cs	58.4	32.0	42.6	–	43.6	34.6	48.9	30.7	30.5	41.6	27.4	44.3	34.5	35.8	26.3	46.5	39.2	45.7	36.5	43.6	41.3	42.9
da	57.6	28.7	44.1	35.7	–	34.3	47.5	27.8	31.6	41.3	24.2	43.8*	29.7	32.9	21.1	48.5	34.3	45.4	33.9	33.0	36.2	47.2
el	59.5	32.4	43.1	37.7	44.5	–	54.0	26.5	29.0	48.3	23.7	49.6	29.0	32.6	23.8	48.9	34.2	52.5	37.2	33.1	36.3	43.3
es	60.0	31.1	42.7	37.5	44.4	39.4	–	25.4	28.5	51.3	24.0	51.7	26.8	30.5	24.6	48.8	33.9	57.3	38.1	31.7	33.9	43.7
et	52.0	24.6	37.3	35.2	37.8	28.2	40.4	–	37.7	33.4	30.9	37.0	35.0	36.9	20.5	41.3	32.0	37.8	28.0	30.6	32.9	37.3
fi	49.3	23.2	36.0	32.0	37.9	27.2	39.7	34.9	–	29.5	27.2	36.6	30.5	32.5	19.4	40.6	28.8	37.5	26.5	27.3	28.2	37.6
fr	64.0	34.5	45.1	39.5	47.4	42.8	60.9	26.7	30.0	–	25.5	56.1	28.3	31.9	25.3	51.6	35.7	61.0	43.8	33.1	35.6	45.8
hu	48.0	24.7	34.3	30.0	33.0	25.5	34.1	29.6	29.4	30.7	–	33.5	29.6	31.9	18.1	36.1	29.8	34.2	25.7	25.6	28.2	30.5
it	61.0	32.1	44.3	38.9	45.8	40.6	57.2	25.0	29.7	52.7	24.2	–	29.4	32.6	24.6	50.5	35.2	56.5	39.3	32.5	34.7	44.3
lt	51.8	27.6	33.9	37.0	36.8	26.5	41.0	34.2	32.0	34.4	28.5	36.8	–	40.1	23.3	38.1	31.6	31.6	31.0	31.8	35.3	35.3
lv	54.0	29.1	35.0	37.8	38.9	29.7	42.7	34.2	32.4	35.6	29.3	38.9	38.4	–		41.5	34.4	39.6	29.3	33.3	37.1	38.0
mt	72.1	32.2	37.2	37.9	38.9	33.7	48.7	26.9	25.8	42.4	22.4	43.7	30.2	33.2	–	44.0	37.1	45.9	38.9	35.8	40.0	41.6
nl	56.9	29.3	46.9	37.0	45.4	35.3	49.7	27.5	29.8	43.4	25.3	44.5	28.6	31.7	22.0	–	32.0	47.7	30.1	30.1	34.6	43.6
pl	60.8	31.5	40.2	44.2	42.1	34.2	46.2	29.2	29.0	40.0	24.5	43.2	33.2	35.6	27.9	44.8	–	44.1	38.2	38.2	39.8	42.1
pt	60.7	31.4	42.9	38.4	42.8	40.2	60.7	26.4	26.2	53.2	23.8	52.8	28.0	31.5	24.8	49.3	34.5	–	39.4	32.1	34.4	43.9
ro	60.8	33.1	38.5	37.8	40.3	35.6	50.4	24.6	26.2	46.5	25.0	44.8	28.4	29.9	28.7	43.0	35.8	48.5	–	31.5	35.1	39.4
sk	60.8	32.6	39.4	48.1	41.0	33.3	46.2	29.8	28.4	39.4	27.4	41.8	33.8	36.7	28.5	43.0	39.0	43.3	35.3	–	42.6	41.8
sl	61.0	33.1	37.9	43.5	42.6	34.0	47.0	31.1	28.8	38.2	25.7	42.3	34.6	37.3	30.0	45.9	38.2	44.1	35.1	38.9	–	42.7
sv	58.5	26.9	41.0	35.6	46.6	33.3	46.6	27.4	30.9	38.9	22.7	42.0	28.2	31.0	23.7	45.6	32.2	44.2	32.7	31.3	33.5	–

Figure 19 Performance obtained with the same standard statistical translation system applied over 22 different European languages. The translation system is based on the standard Moses toolbox, the corpus used is the JRC-Acquis corpus (see chapter 7), and the metric used is the BLEU score. Dark grey cells correspond to a BLEU score performance over 0.5, and light grey cells to a BLEU score performance under 0.4 (blank: between 0.4 and 0.49). Language abbreviations: bg: Bulgarian; cs: Czech; da: Danish; de: German; el: Greek; en: English; es: Spanish; et: Estonian; fi: Finnish; fr: French; gr: Greek; hu: Hungarian; it: Italian; lt: Lithuanian; lv: Latvian; mt: Maltese; nl: Dutch; pl: Polish; pt: Portuguese; ro: Romanian; sk: Slovak; sl: Slovene; sv: Swedish (note that et, fi, and hu are Finno-Ugric, mt is Semitic, and all other languages are Indo-European). Figure taken from Koehn et al., 2009. Reproduced with the authorization of the authors.

translate, since they can add—or "agglutinate"—several morphemes at the end of a lexical form, such as case markers expressing the function of the word in the sentence (along with morphemes expressing possession, determination, etc.). Some Indo-European languages, such as Slavic languages or even German, although not considered as agglutinative, have a rich morphology and do not obtain very good scores. Verbs with a separable prefix and compound words are also difficult to process, which explains why German obtains poor results.

For morphologically-rich languages, a proper syntactic analysis is necessary to provide an accurate translation; for example, one must know if a noun is the subject or the object in order to decide if it is the nominative or the accusative form that should be selected for the translation. Figure 20 also shows some simple examples for Finnish, a

I bought the book.
Minä ostin kirjan.

I did not buy the book.
Minä en ostanut kirjaa.

The book is on the table.
Kirja on pöydällä.

My book is on the table.
Kirjani on pöydällä.

Figure 20 Variations of the word "*book*" in Finnish, depending on its grammatical function.

language in which it is possible to generate an almost infinite number of word forms from a simple word because various morphemes can be added to the basic word form.

Statistical methods can identify correct translations even without undertaking a deep syntactic analysis, especially with the "segment-based approach." A segment being a sequence of words, this approach directly takes advantage of the context (since the context is nothing more than the sequence of words around a given word) and avoids the problems of a purely word-for-word approach. The probability of finding a correct translation, despite everything, becomes inevitably weaker for a morphologically-rich language than for a language where the words vary little as a function of the context, primarily due to a heavy use of prepositions and determiners. Languages like English or French are called "analytical" or "isolating" languages, since they have little variation in terms of morphology and a complex system of prepositions. Chinese is also an analytical language, though Finnish is not, as we have just seen.

For the same reason, it is clear that there is a major bias in the evaluation procedure: evaluation is based on the number of sequences of words shared between an automatic translation and one or several reference translations. Agglutinative languages are thus clearly disadvantaged because, for this kind of language, morphemes are concatenated (i.e., "glued" or "agglutinated") to the basic word forms. The result is that, for these languages, one long word may include several morphemes corresponding to many different types of linguistic information, whereas French or English can deliver the same information merely through a sequence of several small invariable words. This kind of sequence in French or English is then obviously the source of lots of relevant segments ("n-grams") for evaluation. The disadvantage is thus twofold for morphologically-rich languages: analytical languages present more long sequences of words that are likely to improve the evaluation scores (such as frequent prepositional phrases, for example), whereas agglutinative languages present complex linguistic forms that are therefore difficult to analyze accurately.

Let's now turn to the case of English. English has, without doubt, a very poor morphology (especially when compared to other languages), which contributes to the good scores obtained for this language. Most of the time, automatic systems do not have to calculate the right word form in context for English, since words vary little. The availability of very large amounts of data is also a considerable advantage, of course (calling to mind Mercer's "there is no data like more data"; see chapter 7), but, beyond that, it is also the specificities of English, especially its poor morphology, that explain the good results obtained for this language. This naturally brings us to take a look at the errors produced by translation systems.

Typology of Translation Errors

There are very few studies proposing a typology of errors made by machine translation systems. Such a task is in any case difficult and subjective, partly because it depends on the language and on the translation system considered, and partly because the errors are difficult to classify and often vary.

Vilar et al. (2006) tried nonetheless to propose such a typology. Their typology included the following categories: unknown words (words in the source language unknown to the translation system), poorly translated words (wrong meaning, incorrect word form, badly translated idiomatic expression, etc.), word-order problems (problems related to the word order in the target language) and missing words in the target sentence. They show that such an analysis is possible in specific cases (especially when the language pair concerns closely related languages) and can help identify certain weaknesses in the system to resolve them later on (systematic word meaning error, etc.). This type of analysis is especially useful in the case of rule-based systems developed manually, because it allows the developer of the system to correct certain rules or formulate new rules when faced with the main weaknesses observed.

As for statistical systems, the sources of error are more widespread and much more difficult to correct since the systems are not intended to be modified manually. In practice, the system must be "retrained" with new data to have a hope of correcting the identified errors, but the procedure is cumbersome. Moreover, since training is done on very large quantities of data, errors cannot be corrected one by one, and the learning procedure cannot be fully controlled since the process is by definition global and automatic. It is thus hard to correct a specific error in the case of a statistical machine translation. Hybrid systems, as we have seen, try to keep the best of both worlds, making it possible to make generalizations from large amounts of data while keeping as far as possible the ability to make accurate and local corrections.

Finally, it should also be kept in mind that it is indeed the language pair that is the key variable: the types of errors depend, above all else, on the characteristics of each language considered, for the reasons outlined in the previous section (availability of large or small amounts of data for training, morphologically-rich or -poor language, etc.).

Machine translation is sometimes criticized for more fundamental reasons: the techniques used in the field remain to a large extent very close to the text, meaning that the result will also be close to a word-for-word translation (or phrase-for-phrase translation). However, we have seen some cases where a proper translation requires the analysis of the complete sentence and cannot simply be based on local equivalencies between words or phrases; see, for example, the case in chapter 10, where "*the poor don't have any money*" is translated by "*les pauvres sont démunis.*" Kay

(2013) cites the more complex case of *"please take all your belongings with you when you leave the train,"* which corresponds to the French *"veuillez vous assurer que vous n'avez rien oublié dans le train."* These two sentences are semantically similar and are often heard in trains before arriving at a destination. They adopt a different logic, however, with English insisting on the bags to take, and French on the fact of not forgetting anything. Kay considers that this type of translation is extremely frequent and is out of reach of automatic systems. While we may agree with Kay on this last point, it is perhaps not as fundamental as he claims. A correct translation could be found in French that is much closer to the English original sentence, for example: *"veuillez vous assurer que vous prenez tous vos effets avec vous au moment de quitter le train."* This is the type of translation that an automatic system would aim for. More idiomatic translations (even literary ones) are a sign of human translation, but this is not necessarily the goal of machine translation systems.

THE MACHINE TRANSLATION INDUSTRY: BETWEEN PROFESSIONAL AND MASS-MARKET APPLICATIONS

Machine translation is a popular application because it answers a very direct and simple need. Everybody can clearly see the importance of a system that is capable of automatically translating texts from a source language into a target language. It is possible nowadays to access foreign newspapers online without having to master a foreign language. It is even possible to exchange messages on social networks by breaking down language barriers. All this means that economic issues are now important.

A Major Market, Difficult to Assess

The needs and budgets related to translation (by humans or by machines) are unknown and almost impossible to estimate. Companies and public administrations very rarely give information on their translation budgets. Furthermore, the market is extremely fragmented, since translators are often self-employed. Evaluations on the order of several billion dollars (14 to 100 billion) are mentioned here and there, but these estimations are rather unreliable. This is reflected in the considerable variations in published figures.

A Quick Overview of the Market

The European Commission is often cited for its translation budget. This budget is indeed quite large, as some documents must be available in more than 20 languages. According to the Directorate General for Translation,[1] the European Commission's internal translation budget was around 330 million euros in 2013. The number of pages translated has increased to more than two million per year, and more than 93% of these translations are done entirely manually. In fact, according to the same source, less than 5% of translations benefit from automatic aided tools (via the web or internal tools).[2] All of the texts translated for the European Commission are of a technical nature, but the variety of genres and topics addressed is very broad, even if legislative texts predominate. Among the translated documents are technical reports ("white papers") on various topics, correspondence with member states, and websites. In a context such as the European Commission, it is easy to imagine that machine translation could provide valuable services. This is certainly the case when one has to translate specialized and recurring documents, for which the current state-of-the-art technology could give reasonable results. The European Commission has indeed been a longtime investor in machine translation, with Systran particularly, as we have already seen (see below in this chapter for a more detailed presentation of commercial systems, as well as the history of Systran). More recently, the Commission has funded the production of free software and resources, which has allowed for significant advances in the field. Worthy of mention are the parallel corpora Europarl and JRC Acquis (see chapter 7), as well as the development of the software platform Moses[3] thanks to several European funded projects.

Beyond the case of the European Commission, there are a multitude of industrial contexts where technical texts must be translated and updated regularly. Environment Canada's weather forecasts are a good example of such a case, as several versions of the weather forecasts have to be produced daily in both French and English. The production of these bilingual forecasts has been automated since the 1970s with some success (see chapter 6). It is somewhat surprising that this application has been for a long time the flagship application in machine translation; no other iconic system has emerged for other applicative domains.

The production of multilingual leaflets and manuals, as well as localization (i.e., the adaptation of a piece of software for various countries) are important markets for machine translation. Everyone has already experienced trying to understand a leaflet, written in a hard-to-understand manner that was without a doubt the result of an automatic translation. Producing a document (a leaflet, a manual, etc.) in several languages and keeping it up to date comes at a high price, especially for manufacturers selling cheap products and having a low budget for translation. In this context, machine translation is seen as an interesting technology, one that produces texts that can be reviewed by a human translator. Of course, when the process is entirely automatic, with nobody involved to check the results, translations are often of very poor quality.

Another important market for machine translation is access to international patents, which requires specific resources. Patents can be written in a wide variety of languages. Manufacturers launching a new product on the market must ensure that it does not violate a patent in one of the countries concerned. There is thus a specific need to break the language barrier, since a patent in a specific language can be the source of major problems with high financial impact. Another related issue is that patents are written in a very specific jargon. Systems must thus be tuned in order to be able to deal efficiently with the specific terms and phraseology of the domain. What is already a great challenge when dealing with one language only becomes even more crucial and difficult in the context of multilingual systems. Indeed, this field has attracted a great deal of interest, since the commercial and financial profits are high. Large companies are also working on the topic: for example, the European Patent Office is working with Google to propose a machine translation system adapted to the domain. The World Intellectual Property Organization has developed its own translation system based on neural networks in order to translate from Chinese, Japanese, and Korean patent documents into English. Most manufacturers in the field of machine translation offer commercial solutions with regard to the area of patents.

Finally, government intelligence services must be mentioned, as nowadays they are among the largest consumers of machine translation products. This market is little known and even more difficult to assess because, by definition, intelligence services communicate sparingly about their activities. Machine translation has of course to do with the interception of communications. Intelligence agencies cannot analyze all intercepted messages, nor have specialists available for all relevant languages: it is thus crucial to be able to automatically identify the language used in these messages and translate some of them automatically, at least superficially. It is easy to understand why machine translation is a very useful technology for the field, whether applied to written documents or spoken transcripts. Translation needs often concern languages of so-called sensitive countries and often fluctuate according to international affairs. Bearing in mind national interests, the majority of Western countries have developed more or less discrete partnerships with machine translation companies. The ability to produce efficient systems for new languages in a very short time in order to counter new threats quickly and efficiently is one of the key challenges for the domain.

Clearly, the machine translation market is very fragmented, ranging from modules that are freely available on the web (Bing Translator, Google Translate, Systranet, etc.) to purely commercial tools. Additionally, commercial tools are frequently sold in several versions: it is, for example, common to find a free version of a given piece of software on the web along with a professional version sold through more traditional sales channels, most of the time with related services (especially for the development of specialized dictionaries, terminologies, and phraseologies). It should be noted that most companies do not make much money directly from software sales but get most of their revenue through advertisement or services. The services sold around machine translation mean there is still some convergence between technology and professional translators, who are still needed even in this automation context.

Recently the market has also diversified with the growth of speech translation, a field that is still emerging but very promising in terms of concrete applications, particularly on mobile devices. Lastly, we will see that machine translation also provides useful tools for professional translators, even if most automatic tools are not developed directly with this in mind.

Free Online Software

Since the 1990s, several free machine translation systems have emerged on the Internet. One of the first systems, Babelfish, appeared at the end of the 1990s provided by the search engine AltaVista (the most popular search engine at the time). Babelfish was in fact the result of an agreement between AltaVista and Systran, the technology provider for Babelfish. Babelfish was later sold to Yahoo in 2003 and eventually replaced in 2012 by Bing Translator, a product developed and owned by Microsoft.

Today the most well-known free translation service on the Internet is without a doubt Google Translate. Google has conducted research on machine translation since the beginning of the 2000s in order to develop its own solution. The online translator proposed by Google was initially Systran-based, but Systran was gradually removed as Google developed its own technology, first for Russian, Chinese, and Arabic in 2005, for then an online system capable of translating between 25 language pairs in October 2007. Google's system now handles more than 100 languages, with very variable quality depending on the language pair considered.

Google Translate is based on a statistical approach, following the model originally developed by IBM. However, it is clear that these models have since evolved tremendously, even if we don't know the details of the algorithms used, which remain secret. One of the major strengths of Google is that it can rely on its search engine and on its incredible computing power to make the best of the bilingual corpora available on the web. Google's translation system also integrates terminologies and semantic resources when available and has recently begun to deploy a new generation of systems based on the deep learning approach.

Beyond Google Translate, there are plenty of other free automatic translation software packages available on the Internet. As mentioned above, Microsoft's Bing Translator has been adopted by Yahoo to replace its earlier system Babelfish. Meanwhile, Systran offers its own online service called Systranet, and Promt, Systran's main competitor, also offers free translation services online. A multitude of other systems are available directly on the Internet, some specialized in less common languages. A

2010 document compiled by John Hutchins for the European Association for Machine Translation, a compendium of translation software,[4] lists dozens of available products on the Internet. New software and websites appear each week.

Some websites or mobile applications, particularly social networks, also integrate machine translation services to give their users access to content in foreign languages. Facebook and Twitter use Bing Translator to allow end users to access content in foreign languages; recently, Facebook began developing its own in-house technology. Other social networks also integrate machine translation technology. Users can sometimes be unaware that they are reading a machine translation, when this has been displayed automatically, without their intervention (this generally depends on the settings of the social network).

Companies propose these online services for different purposes and get different kinds of revenue from them. For Google and Microsoft, machine translation is considered a key technology in an ecosystem of services aiming to offer better access to information. Machine translation is thus a key component, beyond direct return on investment. Google's main revenue is from advertising, whereas Microsoft receives most of its revenue from software sales (while seeking at the same time to diversify revenue to advertising). For companies such as Systran or Promt, Internet presence is essential, first and foremost to ensure exposure relative to competing products. Advertising and online product sales (including the integration of translation services into other websites, generally generating a revenue proportional to the number of translations per month) is another source of income for software companies.

One can also note that these tools are no longer just standalone online applications. It is now often possible to correct the translations obtained directly online. The system can in turn use these manual corrections to identify some errors and correct itself dynamically, and at no cost, simply by integrating the user's proposed corrections. User feedback is still marginal, but the more a tool has an active community of users, the more it will benefit from this type of feedback. This source of information could prove valuable in the future, especially if automatic approaches reach a plateau (i.e., if improvements slow down after initial rapid progress). In this context, the main source of progress will probably consist in integrating local improvements proposed by users themselves. However, it is generally very difficult and very costly to have access to a community of users, since software customers tend to give little feedback. From this point of view, an online translation service with a large audience is an extremely valuable product.

Finally, online products in no way guarantee the confidentiality of translated data, which are, on the contrary, generally saved and stored by machine translation systems. Most systems keep track of the texts proposed by end users as well as of the proposed translation and use this as a translation memory in which past translations can be found. Thus, companies that need to translate confidential data should by no means use these free products, but should preferably resort to commercial products.

Commercial Products

Along with free products, a multitude of commercial products coexist to respond to various needs and to the different languages represented on the Internet.

Several companies, like Systran and Promt, market solutions for machine translation. Beyond these two companies, many other software companies propose "off-the-shelf" machine translation solutions, sometimes for only a few dollars. These systems are hard to adapt, and their quality is generally quite questionable. This kind of product is now rather marginal and will probably become even more marginal in the future due to the availability of free translation tools online.

A larger market involves the sale of machine translation solutions that can be integrated into websites. We have already seen that Facebook and Twitter first resorted to Bing for translating messages exchanged online, and Facebook is now developing its own "in-house" solution.

Almost all large software integrators propose a machine translation solution that can be integrated into a website. IBM, for example, has developed its own product that is sold as a module in the IBM WebSphere platform. Oracle relies on an agreement with Systran. As we have already seen, the European Patent Office turned to Google and signed agreements with other patent offices in order to improve their machine translation technology (for Chinese, in particular).

Beyond these large and well-known worldwide companies, several other companies propose more focused commercial products for different language pairs. Some regional markets are dominated by local companies, such as Promt in Russia or CSLi in Asia. One can also find companies specialized in specific rare languages or more regional areas. The quality of these systems is highly variable. Moreover, as we saw in the previous chapter, performance is highly dependent on the existence of bilingual parallel corpora and lexicons.

As already said, it should be noted that online sales are generally limited, even if advertising can be a valuable source of additional income. Most of the income for traditional software companies in this domain comes from big companies and large administrations. In this regard, the defense sector is extremely important, especially with the generalization of the interception of communications (via telephones or the Internet). In an interview in a French magazine,[5] the former CEO of Systran, Dimitris Sabatakakis, once said that Systran would not exist without the American intelligence agencies. Indeed, Systran's first revenues were due to an initial contract with the US Army in the 1970s. Systran still benefits today from large contracts with various American defense agencies, as we will now see.

The Case of Systran

The oldest and most well-known company in machine translation is without a doubt Systran (whose name comes from the abbreviation "system translation"). Peter Toma, a researcher who had previously worked at Georgetown University during the early 1960s, founded Systran in 1968. The company initially had American defense organizations (like the US Air Force) as its main customers and was naturally interested in the Russian-English language pair.

The company is also known for having worked with the European community for several years. A demonstration was first carried out in 1975 at the demand of the European Commission. This led to a request for a demonstrator that was subsequently installed in Brussels in 1981. The number of languages covered gradually increased, and this contract ensured regular revenue for Systran. Relations with the European Commission deteriorated when the commission, wishing to part company with the vendor launched a call for bids in 2003 in order to improve the translation system and its dictionaries. Systran filed a lawsuit for copyright violation (on software and related information) and disclosure of confidential data to third parties. Systran finally won its case against the European Commission in 2010.

This lawsuit is not anecdotal. It shows that the quality of a translation system is fundamentally related to the resources it uses, especially for a system relying mostly on dictionaries and rules developed by linguists, as was the case for Systran. In this field, it is crucial to be able to respond quickly to new needs, which means being able to cover new languages and new specialized fields without necessarily having very large corpora available. Indeed, specialized companies like Systran and Promt still primarily offer systems that rely on dictionaries and transfer rules (this was especially the case in the 1980s and 1990s for Systran, before the revolution of the statistical approach in the domain). After the success of Google Translate, Systran developed a "hybrid" approach by adding statistical information to the system, but the basis of the translation model remained relatively traditional, and Systran is now focusing on deep learning, like all the major players in the field. The advantage of keeping a relatively traditional approach is that, even without a training corpus, bilingual dictionaries can be developed, as well as transfer rules from one language to another. Depending on the language pair, it may even be possible to recycle some data for a given language, which is a considerable advantage.

This brings us to the defense market. The CEO of Systran, in the interview previously cited, revealed that a quarter of Systran's revenue during the year 2000 came from US defense industries. The French and Korean markets were also fundamental for the company. We can thus estimate that, in 2000, more than half of the company's revenue related to the defense and intelligence markets (since the US defense industries were not the only defense and intelligence markets where Systran was active). In this context, it is often difficult to have access to training data, as corpora in this domain are highly confidential. Moreover, the world of military and intelligence services wants to be able to adapt a system itself without disclosing data. It is therefore still relevant to work with dictionaries and rules, since it is easy to add new words to an existing dictionary for example, whereas retraining a statistical system is complex and requires large amounts of bilingual data that may not be available. This largely explains why many commercial systems are still based on a traditional approach, using manually developed bilingual dictionaries, even though statistical approaches now dominate the research landscape.

A Worldwide Market

The importance of this strategic market drove large companies in the telecommunications field to strengthen their teams in the field of speech analysis and machine translation. Several company buyouts have taken place recently: Systran was bought in 2014 by a Korean company, CSLi, who developed the voice analysis and translation systems used by Samsung's connected devices (cell phones, tablets, and other technological gadgets). Facebook bought out different companies specialized in machine translation (such as Jibbigo in 2013 for voice messages in particular). Apple and Google are also regularly buying startups in the communication and information technology domains. Most importantly, all these large companies are hiring engineers and researchers (mainly in machine learning and artificial intelligence) in order to produce their own machine translation solution. They are also opening new research centers worldwide in order to attract the best talent everywhere.

New Applications of Machine Translation

The machine translation market is growing fast. Over the last few years we have witnessed the emergence of new applications, particularly on mobile devices. Speech translation has become a hot topic ("speech to speech"

The machine translation market is growing fast. Over the last few years we have witnessed the emergence of new applications, particularly on mobile devices.

applications aim at making it possible to speak in one's own language with another interlocutor speaking in a foreign language by using live automated translation).

Cross-Language Information Retrieval

Cross-language information retrieval aims to give access to documents initially written in different languages. Consider research on patents: when a company seeks to know if an idea or a process has already been patented, it must ensure that its research is exhaustive and covers all parts of the world. It is therefore fundamental to cross the language barrier, for both the query (i.e., the information need expressed through keywords) and the analysis of the responses (i.e., documents relevant to the information need).

A cross-language system is a system that manages multilingualism and accepts queries in a given language so as to identify documents in any languages different from the source language. A machine translation system can then propose a translation of the identified documents into the user's language. This field is the topic of much research at the moment and combines search engines with machine translation to obtain the most accurate result possible.

The main problem is at the level of the information need expressed by the query: Internet queries are, for the most part, composed of one or two keywords, which means that there is too little context to disambiguate keywords. To solve this problem, one possible strategy is to identify (by means of a dictionary) the degree of ambiguity of the words in the query and ask the user to better specify his or her query if necessary (interactively, if efficient strategies are available). An alternative approach involves showing documents answering the query directly (i.e., with no disambiguation stage), before asking the end user to evaluate their relevance according to information need. The automatic analysis of the selected documents can then in turn be used to enhance the search to interactively obtain more accurate results in the target languages. Researchers in this field have developed several products that are primarily integrated in commercial solutions for "key corporate accounts" (large companies or administrations). In order to be efficient, the proposed solutions require the use of specialized lexicons depending on the target field.

Automatic Subtitling and Captioning

Automatic subtitling is an application that automatically produces the transcription of the audio portion of a program. It can be used in a monolingual environment but is also now used to provide live audio translations. Automatic subtitling can often be seen in noisy environments (train stations, airports, etc.) and is already deployed by mass media around the world. This type of application also makes mass media accessible to hearing-impaired people and to people who do not know the source language.

The quality of automatic speech transcription has allowed for these types of applications to exist in monolingual contexts. Today, the techniques used for automatic subtitling coupled with machine translation allow for the production of subtitles in various languages, live and without additional cost. The quality of the result, however, remains a problem, and applicable solutions are not yet deployed on a large scale.

Direct Translation in Multilingual Dialogue

Automatic speech translation is seen as a major opportunity by most information technology companies. Skype, for example, owned by Microsoft, developed a prototype that was incorporated into its communication platform. The trend is now widespread: the mobile messaging application WeChat has also announced the integration of a machine translation system. WeChat is first and foremost an interactive service of written messages, but it also allows for exchanging voice messages: these will be automatically translated in the same ways once the quality of the system is considered sufficient. Finally, Google introduced a voice translation application for mobile devices as part of its Google Translate system on Android platforms.

All mobile operators work on these types of application to allow for "transparent" multilingual calls thanks to a direct translation (that is, to allow calls from callers speaking a foreign language without identifying that the interlocutor is indeed speaking in another language). However, it should be noted that the quality of these systems is unlikely to be sufficient to allow real conversations between humans in the short term: even though the quality of speech transcription improves regularly, the current error rate, combined with different translation modules, risks turning conversation into a dialogue of the deaf!

As for the American giant AT&T, it has developed Watson,[6] a project that enables "speech-to-speech" applications, or live multilingual interaction with simultaneous translation. The application seems more capable of performing successfully than the applications described in the previous paragraph. In fact, in addition to traditional conversations between people, the American company targets audio-interactive services. In this context, translation is highly focused, since the goal of the system is mainly to manage access to large databases of company information. The system must be able to understand a query (expressed in some specific language) and provide an answer (a telephone number, for example) in the speaker's language. This kind of application seems more achievable in the short term than multilingual conversations on any topic between people.

Cell Phones and Connected Objects

New technologies and new applications now play a leading role in machine translation. "Speech-to-speech" applications are inseparable from the development of mobile phones. The majority of the applications we have seen are available today for cell phones (as applications), even if it is not really possible to have a direct multilingual conversation by telephone yet. For practical reasons, mobile phone applications are now geared more toward the direct translation of a few sentences in a conversation between people in the same room for example, but the eventual target is of course the direct translation of distant conversations through mobile phones.

Developers of such applications make use of all the possibilities of modern cell phones. To give one practical example, a specific application makes it possible to take a picture of a restaurant menu and immediately get the translation of the menu (though it seems the system is still unable to say whether the food will be good!). Through this specific application, one can see the convergence of different research fields: image analysis (in order to identify and extract text zones from the image), automatic character recognition, and machine translation.

Internet-connected objects (such as watches or glasses) will serve to support new applications in which multilingual speech will also be included. The Japanese company NTT Docomo has introduced a model of glasses with enhanced vision that incorporates machine translation features: the user can look at a text in Japanese and obtain a translation in English. At the moment, it is just a prototype whose quality and robustness have not been tested, but these examples illustrate the range of applications that exist for both text and speech.

Today these gadgets seem to suffer from a lack of interest from the general public, as a result of their high price and their uncertain positioning in terms of applications (Google Glass generated massive media coverage, only to be pulled from the market due to lack of commercial interest). The future of these objects is without a doubt more promising in professional contexts requiring people to work hands-free, for maintenance in particular (such as in the nuclear, aeronautic, and computer science fields). Other professional contexts could also provide opportunities, such as applications in medicine or sales or in the cultural domain (e.g., visits of museums with augmented reality devices).

Translation Aid Tools

While there has been renewed interest in machine translation since the 2000s, translation aid tools still lag behind. Companies now provide efficient specialized tools, especially "translation memories," or databases where translators can find examples based on previous translations. Translation memories are being increasingly used and sometimes even imposed by companies on translators to ensure the coherence of translations.

Statistical translation models are based on the analysis of large bilingual corpora that can be considered a huge translation memory. However, we must not go too far with the analogy: the work of a human translator has little to do with how automatic systems operate.

Another question is whether machine translation tools, which have made great progress in recent years, can help human translators in their work. Since most tools provide complete translations (and not merely fragments of translations), the only possible strategy consists in post-editing the translation to obtain a quality result. The outcome of this approach is mixed and difficult to generalize. It is necessary for the translations proposed by the machine to be of good quality so that the translator can work quickly and efficiently. The approach is only possible if the system has been tuned to fit the target domain and if the domain has a regular terminology and phraseology. A good example is the system developed for Environment Canada: the target was a very specific field (weather forecasts) with specific pieces of information (temperature lists for each city, etc.) to fill regular text templates. In this context, post-editing is very limited. In comparison, the translation of a technical text with a standard tool risks giving inoperable results.

Machine translation systems sometimes provide a translation that is just sufficient to get the gist of a document. This poor quality, which may be enough in some contexts, is generally very insufficient for a human translator. It also regularly happens that the proposed translation fragments are impossible to reuse. The solution is then to completely rephrase the sentence, and in this case the automatic system is simply useless. Consequently professional translators often prefer traditional work methods, which in the end are faster than automated methods. It is also worth recalling that the European Commission poured a lot of money into machine translation, but that, as mentioned previously, at most 5% of the translations produced were based on automatic tools. This shows that automatic translation is still far from being usable in real-world industrial or administrative contexts, if the target is a nearly professional quality translation.

Machine translation post-editing has, despite everything, recently become a full-fledged field of research. Conferences are currently organized around this single theme, showing the scientific and economic potential of the field. The interest is actually twofold. First, improving the productivity of translators: this involves efficient systems and strategies to make the best of the output of machine translation tools. Second, improving machine translation systems directly: this means being able to dynamically reuse end-user feedback to make the system evolve and propose more accurate translations in the future.

Beyond these experiments using standard machine translation tools, there is broad consensus today that translation aid tools should not supply a single translation at sentence level, but fragments of translations from which the translator can then choose. Trojanskij's assisted translation environment (see chapter 4) remains in this regard a clear-sighted invention that has still never been explored in depth. We may also recall Bar-Hillel's recommendations, or the 1966 ALPAC report: high-quality machine translation was seen as an illusion or at least an elusive goal for a long time. In the meantime, human translators need specific tools (and not standard commercial machine translation systems) to improve their productivity as well as the quality and homogeneity of the translations produced.

This is in fact a difficult problem, since no one knows exactly what would truly be helpful for a translator. Enhancing translation memories is the easiest path, since it displays the most relevant segments of texts according to the context of translation. But even this seemingly modest enhancement poses problems, insofar as continuously updating the displayed translation fragments can make the application relatively slow and burdensome to use. However, translation memory modules are widely used and remain the main application employed by professional translators in their work environments.

CONCLUSION: THE FUTURE OF MACHINE TRANSLATION

Throughout the overview given in this book, we have seen the evolution of machine translation, from the first experiments in the 1950s to today's systems, which are operational and available on the Internet at no cost. We also saw the main features of these systems: some based on dictionaries and transfer rules, others on the statistical analysis of very large corpora. Lastly, we have described a new approach based on deep learning that seems highly promising. This new approach is especially exciting from a technical and cognitive point of view. But let us first have a look at commercial challenges.

Commercial Challenges

As previously noted, machine translation has undergone a profound renewal since the 1990s, the period when very large amounts of bilingual documents became freely and directly available online. At the same time, the development of the Internet played a crucial role, since people can now communicate worldwide through email, blogs, and social networks. There is thus a need for tools making it possible to communicate in different languages, without mastering many languages. This technological revival was therefore supported by commercial and strategic prospects, notably in the fields of telecommunications and information technology.

In the course of this book, we have described several kinds of applications. Everybody is familiar with the translation tools freely available on the Internet and created by Internet giants such as Google or Microsoft. In a multipolar, multilingual world, mastering this technology is an absolute must for Internet and telecommunications companies with global ambitions. Machine translation is the key to multiple products with great potential in the near future, such as live multilingual communication or multilingual access to patent databases. Some of these applications will generate significant revenue in the years to come.

Another type of application is probably not as well known: several specialized companies supply professional commercial translation products. These products are complex, adaptable, and often sold with specific services (especially the development of specialized dictionaries or the rapid integration of new languages on demand). This kind of product is primarily sold to large companies and government administrations, especially in the military and intelligence domains. Strategic interests thus play a major role in this context. The adaptability of such systems is also a key element: a solution provider's ability to rapidly supply accurate translations for a new field or a new language is of utmost importance. In this framework, it is also often crucial for customers to be able to develop resources themselves, especially when the data to be analyzed are classified.

The development of communications networks, mobile Internet, and the miniaturization of electronic devices also highlight the need to switch quickly to audio applications that are able to translate speech directly. Speech processing has been the subject of intensive research in recent decades, and performance is now acceptable. However, the task remains difficult since speech processing as well as machine translation have to be performed in real time, and errors are cumulative (i.e., if a word has not been properly analyzed by the speech recognition system, it will not be properly translated). Large companies producing connected tools (Apple, Google, Microsoft, or Samsung, to name a few) develop their own solutions and regularly buy start-ups in technological domains. They need to be first on the technological front and propose new features that may be an important source of revenue in the future.

The future will likely see the integration of machine translation modules in new kinds of appliances, as seen in chapter 14. Microsoft has already presented live demonstrations of multilingual conversations, integrating speech translation into Skype. Google, Samsung, and Apple are creating similar applications for mobile phones, and even for "smart" eyeglasses. While it is not yet clear whether these gadgets will really be used in everyday life, they are interesting for specific professional contexts, such as the maintenance of complex systems in the aeronautic or nuclear industry, where technicians must be able to communicate while keeping their hands free. It is clear that commercial challenges will continue to drive research toward more powerful and accurate systems.

We live in a multilingual world, yet there are problems of language domination related to machine translation (and to the field of natural language processing as a whole), since the domain is of course not independent of economic and political considerations. As has been emphasized, even if the systems available on the Internet officially propose to translate into several tens of languages, the quality is very poor for most, especially if English is not the source language, or, better, the target language. Aside from Indo-European languages (English, Russian, French, German, etc.), some languages (such as Arabic or Chinese) are now the focus of intensive research. These are usually the most widely spoken languages in the world and associated with great economic potential. One can also find research projects addressing rarer languages, but they remain marginal, and the quality of these systems is generally very moderate. Processing rarer languages remains a highly interesting challenge, as long as it is not dominated purely by economic interests.

A Cognitively Sound Approach to Machine Translation?

To conclude this journey, we would like to say a few words about cognitive issues. The most active researchers in the field of machine translation generally avoid addressing cognitive issues and make few parallels with the way humans perform a translation. The artificial intelligence domain has suffered from spectacular and inflated claims too much in the past, and in relation to systems that had nothing to do with the way humans think or reason. It may thus seem reasonable to focus on technological issues and leave any parallel with human behavior aside, especially because we do not, in fact, know much about the way the human brain works.

However, it may be interesting in this conclusion to have a look at cognitive issues despite what has just been said, because the evolution of the field of machine translation is arguably highly relevant from this point of view. The first systems were based on dictionaries and rules and on the assumption that it was necessary to encode all kinds of knowledge in the source and target languages in order to produce a relevant translation. This approach largely failed because information is often partial and sometimes contradictory, and knowledge is contextual and fuzzy. Moreover, no one really knows what knowledge is, or where it begins and where it ends. In other words, developing an efficient system of rules for machine translation cannot be carried out efficiently by humans, since the task is potentially infinite and it is not clear what should be encoded in practice.

Statistical systems then seemed like a good solution, since these systems are able to efficiently calculate complex contextual representations for thousands of words and expressions. This is something the brain probably does in a very different way, but nevertheless very efficiently: we have seen in chapter 2 that any language is full of ambiguities (cf. "*the bank of a river*" vs. "*the bank that lends money*"). Humans are not bothered at all by these ambiguities: most of the time we choose the right meaning in context without even considering the other meanings. In "*I went to the bank to negotiate a mortgage,*" it is clear that the word "*bank*" refers to the lending institution, and the fact that there is another meaning for "*bank*" is simply ignored by most humans. A computer still has to consider all options, but at least statistical systems offer interesting and efficient ways to model word senses based on the context of use.

We have also witnessed rapid progress, from the very first systems based on a word-for-word approach to segment-based approaches, which means that gradually longer sequences of text have been taken into account, leading to better translations. The new generation of systems based on deep learning takes into account the whole sentence as the basic translation unit and thus offer a valuable solution to the limitations of previous approaches. We have also seen that this approach takes into account all kinds of relations between words in the sentence, which means that structural knowledge (i.e., some kind of syntax) is involved in the translation process. The fact that all this information is embedded and processed at the same time in a unique learning process means that one does not need to deal either with the delicate integration of various complex modules or with the propagation of analysis errors, contrary to what happened with most previous systems (but note that errors can also be percolated into the neural network; the sole use of neural networks does not solve all problems magically, of course).

In practice, deep learning systems still suffer from important limitations, and we saw in chapter 12 a number of the research issues at stake (unknown words, long sentences, optimization problems, etc.). While we are still far

The new generation of systems based on deep learning directly takes into account the whole sentence as the basic translation unit and thus offers a valuable answer to the limitations of previous approaches.

from perfect machine translation systems, it is nonetheless interesting to see that the best-performing systems now operate directly at sentence level, make limited use of manually defined syntactic or semantic knowledge, and produce translations on the fly based on huge quantities of data used for training. They thus seem to account for several characteristics of human language: for example, the fact that child language acquisition is based on language exposure (and not on the explicit learning of grammar rules); and the fact that word distribution and linguistic complexity play a role (some words are more frequent than others and are learned before others, and simpler syntactic structures are easier to acquire and easier to translate). It is not completely clear how neural networks work, what knowledge they effectively use, and how their architecture influences the overall result, but it is clear that they bear interesting similarities to basic features of human languages.

As already said, deep learning machine translation is still in its infancy. We can expect quick progress as the systems achieve better quality and will gradually appear in a broader number of professional contexts. Automatic systems will, of course, not replace human translation—this is neither a goal nor a desired outcome—but they will help millions of people have access to information they could not grasp otherwise. Digital communication will continue to grow, as will research in the machine translation domain, and one can expect that in the not-too-distant future, it will be possible to dialogue over the phone with someone speaking another language. One will then just need to introduce a small device into one's ear to understand any language, and Douglas Adams' Babel fish will no longer be a fiction—although the device may not be a fish!

NOTES

Chapter 1

1. *The Hitchhiker's Guide to the Galaxy* was originally a radio comedy broadcast (1978) before giving birth to different adaptations, including comics, novels, TV series, and plays.

2. Babelfish is also the name of a machine translation system that was very popular on the web in the late 1990s.

3. Alan Turing was a British mathematician, logician, and computer scientist. He played a major role in the development of computer science, and his life has recently been popularized in the movie *The Imitation Game* (2014).

Chapter 2

1. One should note however that very recent advances in the field, based on deep learning, try to avoid translating isolated groups of words and consider instead the whole sentence directly.

2. https://wordnet.princeton.edu.

3. Advertising often plays with ambiguity and double meaning (for example in a slogan like "*Trust Sleepy's, for the rest of your life*," where "*rest*" refers both to the act of resting and to what remains of your life). Most people will not immediately see the double meaning, which means that humans are naturally prone to select one interpretation and not even consider alternate solutions.

Chapter 4

1. The subject is tackled in the correspondence between Descartes and Mersenne, where Descartes explains the problems but also the advantages that would result from such an invention.

2. Hutchins, 1986, chapter 2 ("Precursors and pioneers").

3. *Zur mechanischen Sprachübersetzung: ein Programmierung Versuch aus dem Jahre 1661.*

Chapter 5

1. Weaver, letter to Wiener, March 4, 1947.

2. Wiener's response to Weaver's letter, April 30, 1947.

3. Weaver worked at the Rockefeller Foundation, where he was responsible for launching new research projects.

4. "Thus may it be true that the way to translate from Chinese to Arabic, or from Russian to Portuguese, is not to attempt the direct route, shouting from tower to tower. Perhaps the way is to descend, from each language, down to the common base of human communication—the real but as yet undiscovered universal language—and then re-emerge by whatever particular route is convenient" (Weaver, "Translation," 1955, 23).

5. The fact that most words can belong to several categories, such as the word "*bank*," which can be a noun or a verb, also poses an important problem for automatic systems. The correct analysis of a given sentence requires at the very least a proper recognition of the main verb, since it is the verb that structures the whole sentence. But even this is not a trivial task for a computer!

6. A translation memory is a database that contains previously translated fragments of texts in order to help professional translators quickly find equivalences, while ensuring more regular and consistent translations.

7. "The model ... was ... too crude and has to be replaced by a much more complex but also much better fitting model of linguistic structure" (Bar-Hillel, 1959, Annex II, p. 8).

Chapter 6

1. Specifically the Department of Defense, National Science Foundation, and the Central Intelligence Agency.

2. For example, https://www.nap.edu/openbook.php?record_id=9547.

3. "Several well-known studies indicate that in 200 hours or less a scientist can acquire an adequate reading knowledge of Russian for material in his field" (ALPAC Report, 1966, p. 5).

4. "In this context, there has been no machine translation of general scientific text and none is in immediate prospect" (ALPAC Report, 1966, p. 19).

5. "For those concerned with a better understanding of the structure of human language and, incidentally, the structure of artificial languages such as those used in computer programming, there can be considerable satisfaction in the extensive progress made in the past decade and a good deal of optimism for the future" (Oettinger, 1963, p. 27).

Chapter 7

1. Mercer was a researcher in the IBM research team that founded statistical translation. See chapter 9.

Chapter 8

1. This is known in English as the *marker hypothesis* (Green, 1979).

Chapter 11

1. This example first appeared in a critical press article on machine translation in the early 1960s, but it is not the result of a real translation system; the systems at the time only had available to them dictionaries that were too limited to result in this type of error and, in practical terms, it was actually more their lack of coverage than their inaccuracy that was a problem.

2. https://fkaplan.wordpress.com/2014/11/15/langlais-comme-langue-pivot-ou-limperialisme-linguistique-cache-de-google-translate/.

3. See, for example, Ken Church's article "A pendulum swung too far." The appeal of statistical methods since the 1990s has largely deterred researchers from investigating more fundamental aspects of language that would require a deep analysis.

Chapter 12

1. This is what happens in practice, anyway. The question is then, in fact, whether the structure inferred by the computer makes more sense than the syntactic structure a human would provide.

Chapter 13

1. The Advanced Research Projects Agency (ARPA) is an American agency founded in 1958 and responsible for the development of emerging technologies in the USA. The name of the agency has changed several times, and the agency is probably better known now under the acronym DARPA (where D stands for Defense), its name since 1972 (except between 1993 and 1996).

2. This can also be seen as somewhat contradictory: if the approach is designed as needing to be interactive from the beginning, it is not necessarily relevant to evaluate the translations produced completely automatically. It is, rather, the capacity of the system to provide relevant translational elements that should be evaluated.

Chapter 14

1. Data available online (site visited May 20, 2016); see http://ec.europa.eu/dgs/translation/faq/index_en.htm#faq_4/.

2. The rest is marginal and corresponds to work such as post-editing, translations of summaries, etc.

3. Moses is an open system for machine translation that implements some of the main algorithms of statistical machine translation. Moses incorporates another tool, Giza++, which implements various IBM models, and plenty of

other algorithms have since been included in this platform, which is free online (http://www.statmt.org/moses).

4. http://www.hutchinesweb.me.uk/Compendium-16.pdf; site visited September 15, 2014.

5. In an article from the magazine *Le Point*, September 2013; see http://www.lepoint.fr/editos-du-point/jean-guisnel/dimitris-sabatakakis-systran-n-existerait-pas-sans-les-agences-de-renseignements-americaines-18-09-2013-1732865_53.php.

6. See http://www.research.att.com/projects/WATSON/.

GLOSSARY

Agglutinative language
language in which most of the grammatical information is expressed through **suffixes** added to words. Agglutinative languages are morphologically complex and require an efficient morphological analyzer in order to be processed accurately.

Ambiguity
a word (or any other linguistic unit) that has different meanings. For example "*bank*" can be a money-lending institution or the edge of a river. Ambiguity is pervasive in languages and is one of the main problems natural language processing has to face.

Cognate
related words with a similar form and meaning across languages. Proper nouns are often valid cognates ("*Paris*" designates the same city in English and in French, while "*Londres*" and "*London*" do not have exactly the same form in the two languages but can nevertheless be considered as valid cognates). In contrast, French "*achèvement*" is not a valid cognate of English "*achievement*" since the two words, although etymologically related, do not have the same meaning today (French "*achèvement*" means "*completion*"). This is known as a "deceptive cognate."

Compound
a word composed of several morphemes that generally does not fully preserve the semantics of its components. For example, "*round table*" generally designates an event, most of the time with no "round" table. When a compound is made of several words, it is called a "multiword expression" (as opposed to "solid compounds" like "*football*" or "*blackboard*" where the concatenated morphemes in the end produce one single word unit).

Conjugation
the different forms of a verb obtained by inflection.

Entry, dictionary entry
short description of a given word meaning in a dictionary. Generally if a word has different meanings (i.e. if the word is ambiguous), it has different entries (one entry per meaning)

FAHQT
Fully automated high-quality translation; also found as FAHQMT, for fully automated high-quality machine translation.

Frozen expression
see **idiom**.

Grammatical function
role of the word in the sentence (e.g., subject, object, etc.).

Idiom
complex expression whose meaning has little to do with the semantics of its parts (e.g., "*kick the bucket*," which has nothing to do with "kick" or with "bucket").

Inflection
variation of a given word depending on its grammatical function in a sentence. The term inflection is used for nouns and adjectives. For verbs, people generally use the term **conjugation**, but both terms refer to fundamentally the same process. The more variation there is, the more the language at stake will be called "morphologically complex." English is known to be simpler than many other languages from a morphological point of view.

Interlingua
representation of the semantic content of a sentence in a language-independent formalism.

Lemma
normalized form of a word as found in a dictionary (e.g., "walk" as opposed to "walking").

Lemmatizer
automatic tool intended to calculate the lemma of each word in a text. The task is not obvious when a surface form corresponds to different possible lemma and should thus be disambiguated according to the local context.

Light verb
a verb used in a context where it has little semantic content, especially in complex verbal expressions such as "to take a shower" (where nobody literally takes anything).

Morpheme
word part. See **morphology**.

Morphological analyzer
automatic tool calculating the structure of a word (see **morphology**).

Morphology
analysis of the structure of words. Words are generally made of a stem, with (optionally) some prefixes and some suffixes. For example, in the noun "*deconstruction*," "*de-*" is a prefix and "*-tion*" is a suffix. The word stem is "*construct*" or even "*-struct*" (since "*con-*" can also be considered a prefix). Word parts (stems, prefixes, and suffixes) are called morphemes.

Morphosyntax
see **part-of-speech tagger**.

Occurrence
the presence of a word in a corpus: the number of occurrences of a word in a text is the number of times the word is used in the text.

Parser
see **syntactic analyzer**.

Part-of-speech (or morphosyntax) tagger
automatic tool assigning part-of-speech tags to words in context (e.g., a specific sentence). The task is difficult since most words are ambiguous (e.g., "*fly*" can be a noun or a verb.).

Part-of-speech tags
word categories (noun, verb, adjective, etc.). In English, researchers generally consider around a dozen categories, but the inventory varies greatly across languages.

Phrase
semiautonomous group of words in a sentence, such as a noun phrase (*"a cat"*) or a verb phrase (*"to go shopping"*). A phrase is said to be semiautonomous since it does not form a full sentence in itself, but it can be associated with some autonomous meaning (as opposed to a sequence like *"cat goes to"*).

Precision
fraction of retrieved "information nugget" (words, sequences of words, documents) that are relevant to a query or a task.

Prefix
see **morphology**.

Recall
fraction of the retrieved "information nuggets" (words, sequences of words, documents) that are relevant to a query or a task, and that are successfully retrieved.

Semantic analyzer
automatic tool intended to provide a semantic representation (see **semantics**).

Semantics
analysis of the meaning of any linguistic unit (word, phrase, sentence, or any higher-level unit, such as a paragraph or text).

Suffix
see **morphology**.

Syntactic analyzer or parser
automatic tool intended to provide the syntactic representation of a linguistic unit (see **syntax**).

Surface form or word form
word as it occurs in a text. The proper analysis of surface forms (recognizing the different morphemes and linking a word form to the corresponding **lemma**) has a lot to do with **morphology**. The task is relatively simple for English, which has a relatively low level of morphological complexity (e.g., *"dancing,"* *"dances,"* and *"danced"* are easily recognizable forms of *"to dance"*). Linguists consider that French is morphologically more complex than English

(since there are more surface forms per lemma in French than in English) and Finnish is even more complex (there could even be theoretically a near-infinite number of word forms for one lemma in Finnish since Finnish is an **agglutinative language**).

Syntactic structure
structure of a group of words reflecting their relative grammatical function.

Syntax
structure of a group of words, generally a sentence. The result of a syntactic analysis is generally a tree, in which everything depends on the main verb.

Transfer rule
in a rule-based machine translation system, a transfer rule formalizes the way a linguistic structure in the source language must be rendered in the target language. Transfer rules have to do with syntax.

Vague, vagueness
refers to the fact that a language is never completely precise or could always be more precise, especially in relation to the external world. Vagueness is pervasive in language and involves many differing notions (e.g., vague concepts such as "*to be bald*"; philosophical and abstract concepts such as "*to be good*"; concepts that vary across languages like colors; etc.).

Word sense
different meanings of a word. The number of word senses corresponds to the number of entries for one word in a dictionary.

BIBLIOGRAPHY AND FURTHER READING

This book is accompanied by the following website: http://lattice.cnrs.fr/ma-chinetranslation. The website provides a variety of supplementary material, including corrections of mistakes and other resources that should be useful to readers. This website also presents the analysis of the output of different machine translation systems that could not be included in this book since systems' performance evolves too quickly. In addition, this chapter contains suggestions for further reading for the reader who would like to know more than what could be said in this short introduction. This list is by no means exhaustive, which is in any case an impossible task since new publications appear every day on the topic. The following references can be considered the main ones to consult in order to explore different aspects of the topic in greater detail. Most references contain, in turn, their own list of references on specific issues.

Historical aspects of the topic are very well documented thanks to the comprehensive work of John Hutchins. Some other aspects are more difficult to explore, because they are plentiful and very technical (especially concerning current lines of research) or rare and quickly obsolete (e.g., questions related to the commercial aspects of the field).

On historical aspects, the reader should refer to John Hutchins' website: http://www.hutchinsweb.me.uk. John Hutchins has also written three major books on the question:

John Hutchins (1986). *Machine Translation: Past, Present, Future*. Series in Computers and Their Applications. Chichester, UK: Ellis Horwood.

John Hutchins and Harold L. Somers (1992). *An Introduction to Machine Translation*. London: Academic Press.

John Hutchins (ed.) (2000). *Early Years in Machine Translation: Memoirs and Biographies of Pioneers*. Amsterdam: John Benjamins.

The 1992 book, co-authored with Harold Somers, remains interesting, even if it is of course now dated. The other two books are two musts for anyone interested in the history of machine translation. The 1986 book contains a description of the main research groups involved in the domain up to the early 1980s. It also includes descriptions of the main systems and the techniques used by different research groups. The 2000 book contains more historical anecdotes and personal stories, as well as firsthand accounts by the main actors of the domain.

Other publications by John Hutchins are also interesting to get a quick but reliable overview. For example:

John Hutchins (2010). "Machine translation: A concise history." *Journal of Translation Studies* 13 (1–2): 29–70. Special issue: The teaching of computer-aided translation, ed. Chan Sin Wai.

On corpus alignment and the statistical approach to machine translation, the following books are quite technical but important:

Philipp Koehn (2009). *Statistical Machine Translation*. Cambridge: Cambridge University Press.

Jorg Tiedemann (2011). *Bitext Alignment*. San Rafael, CA: Morgan and Claypool Publishers.

On natural language processing in general, several good overviews exist, for instance:

Dan Jurafsky and James H. Martin (2016). *Speech and Language Processing* (3rd ed. draft). Available online: https://web.stanford.edu/~jurafsky/slp3/.

In what follows, we also give other references that could be useful to the reader who wants to know more on some specific aspects of the question. The following references have also been used as the main sources for this book.

Chapter 2: The Trouble with Translation

Many publications have addressed the problem of translation, but they cannot all be listed here. The recent book by David Bellos is an entertaining and captivating introduction, even if many others could also have been cited.

David Bellos (2011). *Is That a Fish in Your Ear? Translation and the Meaning of Everything*. London: Penguin/Particular Books.

Adam Kilgarriff (2006). "Word senses." In *Word Sense Disambiguation: Algorithms and Applications* (E. Agirre and P. Edmonds, eds.). Dordrecht: Springer.

Chapter 3: A Quick Overview of the Evolution of Machine Translation

This chapter presents a quick overview of the field. The reader should thus refer to the general references given at the beginning of this section.

Chapter 4: Before the Advent of Computers...

The literature on universal languages is huge but the introduction by Umberto Eco is accessible and entertaining. Hutchins' article on Artsrouni and Trojanskij is the main source in English on these two researchers.

René Descartes (1991). *The Philosophical Writings of Descartes*. Volume 3: The Correspondence. Cambridge: Cambridge University Press.

Umberto Eco (1997). *The Search for the Perfect Language*. Oxford: Wiley.

John Hutchins (2004). "Two precursors of machine translation: Artsrouni and Trojanskij." *International Journal of Translation* 16 (1): 11–31.

Philip P. Wiener (ed., 1951). *Leibniz Selections*. New York: Simon and Schuster.

Chapter 5: The Beginnings of Machine Translation: The First Rule-Based Systems

For this chapter, apart from the writings from Weaver and Bar-Hillel themselves, one can refer to Hutchins' text on Weaver ("Warren Weaver and the launching of MT: Brief biographical note") and Y. Bar-Hillel ("Yehoshua Bar-Hillel: A philosopher's contribution to machine translation"), both in *Early Years in Machine Translation* (see the full reference at the beginning of this chapter).

Yehoshua Bar-Hillel (1958 [1961]). "Some linguistic obstacles to machine translation." *Proceedings of the Second International Congress on Cybernetics* (Namur, 1958), 197–207, 1961 (reprinted as Appendix II in Bar-Hillel 1959).

Yehoshua Bar-Hillel (1959). "Report on the state of machine translation in the United States and Great Britain." Technical report, 15 February 1959. Jerusalem: Hebrew University.

Yehoshua Bar-Hillel (1960). "The present status of automatic translation of languages." *Advances in Computers* 1: 91–163.

Richard H. Richens (1956). "A general program for mechanical translation between two languages via an algebraic interlingua." *Mechanical Translation*, 3(2): 37.

Karen Sparck Jones (2000). "R. H. Richens: Translation in the NUDE." In *Early Years in Machine Translation* (W. J. Hutchins, ed.). Amsterdam: John Benjamins, 263–278.

Warren Weaver (1949 [1955]). "Translation." Reproduced in *Machine Translation of Languages* (W. N. Locke and D. A. Booth, eds.). Cambridge, MA: MIT Press, 15–23.

Chapter 6: The 1966 ALPAC Report and Its Consequences

The ALPAC report and some comments on it, especially by John Hutchins, can easily be found on the Internet.

The Automatic Language Processing Advisory Committee (1966). "Language and Machines—Computers in Translation and Linguistics." Washington, DC: National Academy of Sciences, National Research Council. [This publication is more popular under the name "ALPAC Report."]

John Hutchins (2003). "ALPAC: The (in)famous report." In *Readings in Machine Translation* (S. Nirenburg, H. L. Somers, Y. Wilks, eds.), 131–135. Cambridge, MA: MIT Press.

John Hutchins (1988). "Recent developments in machine translation: A review of the last five years." In *New Directions in Machine Translation: Conference Proceedings, Budapest 18–19 August 1988* (D. Maxwell, K. Schubert, and T. Witkam, eds.), 7–62. Foris Publications (Distributed Language Translation 4), Dordrecht.

Anthony G. Oettinger (1963). "The state of the art of automatic language translation: an appraisal" In *Beiträge zur Sprachkunde und Informationsverarbeitung*, n°2, 17–29.

Chapter 7: Parallel Corpora and Sentence Alignment

The book by Tiedemann, *Bi-text Alignment* (Morgan and Claypool Publishers, 2011; see full reference at the beginning of this chapter) gives a general overview on the topic. A few historical research papers remain the main contribution to the domain. See, for example:

William A. Gale and Kenneth W. Church (1993). "A program for aligning sentences in bilingual corpora." *Journal of Computational Linguistics* 19 (1): 75–102.

Martin Kay and Martin Röscheisen (1993). "Text-translation alignment." *Journal of Computational Linguistics* 19 (1): 121–142.

Chapter 8: Example-Based Machine Translation

Several research papers are accessible and give a good overview of the benefits but also the limitations of this paradigm.

Makoto Nagao (1984). "A framework of a mechanical translation between Japanese and English by analogy principle." In *Artificial and Human Intelligence* (A. Elithorn and R. Banerji, eds.). Elsevier Science Publishers, Amsterdam.

Eiichiro Sumita and Hitoshi Iida (1991). "Experiments and prospects of example-based machine translation." *Proceedings of the Twenty-Ninth Conference of the Association for Computational Linguistics*, 185–192. Berkeley, CA.

Thomas R. Green (1979). "The necessity of syntax markers: Two experiments with artificial languages." *Verbal Learning and Verbal Behavior* 18: 481–496.

Harold Somers (1999). "Example-based machine translation." *Machine translation* 14 (2): 113–157.

Nano Gough and Andy Way (2004). "Robust large-scale EBMT with marker-based segmentation." *Proceedings of the Tenth International Conference on Theoretical and Methodological Issues in Machine Translation*, 95–104. Baltimore, MD.

Chapter 9: Statistical Machine Translation and Word Alignment

The most important references (by Koehn on statistical machine translation and by Tiedemann on corpus alignment) were given at the beginning of this chapter. The series of historical papers published by the IBM team in the late 1980s and beginning of the 1990s should be read carefully by anyone interested in statistical machine translation.

Peter Brown, John Cocke, Stephen Della Pietra, Vincent Della Pietra, Frederick Jelinek, Robert Mercer, and Paul Roossin (1988). "A statistical approach to language translation." In *Proceedings of the Twelfth Conference on Computational Linguistics*, Vol. 1, 71–76. Association for Computational Linguistics, Stroudsburg, PA. http://dx.doi.org/10.3115/991635.991651/.

Peter F. Brown, John Cocke, Stephen A. Della Pietra, Vincent J. Della Pietra, Frederick Jelinek, John D. Lafferty, Robert L. Mercer, and Paul S. Roossin (1990). "A statistical approach to machine translation." *Computational Linguistics* 16 (2): 79–85.

Peter F. Brown, Vincent J. Della Pietra, Stephen A. Della Pietra, and Robert L. Mercer (1993). "The mathematics of statistical machine translation: Parameter estimation." *Computational Linguistics* 19 (2): 263–311.

A website (http://www.statmt.org) gives access to a large amount of information on the domain, including research papers, tutorials, links to free software, and so on.

Chapter 10: Segment-Based Machine Translation

The previous website (http://www.statmt.org) is probably the best source of information for recent trends related to statistical machine translation, of which segment-based machine translation is part.

Chapter 11: Challenges and Limitations of Statistical Machine Translation

See http://www.statmt.org,as for chapter 10 above.

Kenneth Church (2011). "A pendulum swung too far." *Linguistic Issues in Language Technology*, 6(5).

Chapter 12: Deep Learning Machine Translation

The book by Goodfellow et al., although technical, offers an affordable and comprehensible introduction to deep learning. One can also refer to the blogs of commercial systems that offer interesting overviews (see, for example, Google's Research Blog, https://research.googleblog.com/2016/09/a-neural-network-for-machine.html, or the Systran blog: http://blog.systransoft.com/how-does-neural-machine-translation-work). Google's paper describing their first operational deep learning machine translation system is also worth being read.

Ian Goodfellow, Yoshua Bengio and Aaron Courville (2016). *Deep Learning*. Cambridge, MA: MIT Press.

Yonghui Wu, et al. (2016). "Google's neural machine translation system: Bridging the gap between human and machine translation." Published online. arXiv:1609.08144.

Chapter 13: The Evaluation of Machine Translation Systems

The BLEU, NIST, and METEOR measures are described in the following three publications:

Kishore Papineni, Salim Roukos, Todd Ward, and Wei-Jing Zhu (2002). "BLEU: A method for automatic evaluation of machine translation." *Fortieth Annual Meeting of the Association for Computational Linguistics*, 311–318. Philadelphia.

George Doddington (2002). "Automatic evaluation of machine translation quality using n-gram cooccurrence statistics." *Proceedings of the Human Language Technology Conference*, 128–132. San Diego.

Satanjeev Banerjee and Alon Lavie (2005). "METEOR: An automatic metric for MT evaluation with improved correlation with human judgments." *Proceedings of Workshop on Intrinsic and Extrinsic Evaluation Measures for MT and/or Summarization at the Forty-Third Annual Meeting of the Association of Computational Linguistics*. Ann Arbor, MI.

We have also cited the following four references:

Martin Kay (2013). "Putting linguistics back into computational linguistics." Conference given at the Ecole normale supérieure, Paris. http://savoirs.ens.fr/expose.php?id=1291/.

Philipp Koehn, Alexandra Birch, and Ralf Steinberger (2009). "462 machine translation systems for Europe." *Proceedings of MT Summit XII*, 65–72. Ottawa, Canada.

David Vilar, Jia Xu, Luis Fernando D'Haro, and Hermann Ney (2006). "Error analysis of machine translation output." *Proceedings of the Language Resource and Evaluation Conference*, 697–702. Genoa, Italy.

John S. White, Theresa O'Connell, and Francis O'Mara. (1994). "The ARPA MT evaluation methodologies: Evolution, lessons, and future approaches." *Proceedings of the 1994 Conference*, Association for Machine Translation in the Americas, 193–205. Columbia, MD.

Chapter 14: The Machine Translation Industry: Between Professional and Mass-Market Applications

There are very few studies on this topic. The Directorate-General for Translation of the European Commission gives some figures on its website: http://ec.europa.eu/dgs/translation/faq/index_en.htm#faq_4/.

The following compendium of companies in the field was quite comprehensive in 2010 but is already obsolete because the field evolves very quickly.

John Hutchins, on behalf of the European Association for Machine translation, (2010). "Compendium of translation software." http://www.hutchinsweb.me.uk/Compendium.htm.

In addition, specialized journals and magazines in computer sciences and information technology report the main news concerning the domain, along with traditional newspapers from the financial domain.

INDEX

THIERRY POIBEAU is director of research at the Centre National de la Recherche Scientifique in Paris and head of the LATTICE laboratory (Langues, Textes, Traitements informatiques, et Cognition). He is also an affiliated lecturer in the Department of Theoretical and Applied Linguistics at the University of Cambridge.